Lóegaire Humphrey (Ed.)

Nokia C2-02

W0259605

Lóegaire Humphrey (Ed.)

Nokia C2-02

Nokia, Touchscreen, Series 40

Claud Press

Imprint

Permission is granted to copy, distribute and/or modify this document under the terms of the GNU Free Documentation License, Version 1.2 or any later version published by the Free Software Foundation; with no Invariant Sections, with the Front-Cover Texts, and with the Back- Cover Texts. A copy of the license is included in the section entitled "GNU Free Documentation License".

All parts of this book are extracted from Wikipedia, the free encyclopedia (www.wikipedia.org).

You can get detailed informations about the authors of this collection of articles at the end of this book. The editors (Ed.) of this book are no authors. They have not modified or extended the original texts.

Pictures published in this book can be under different licences than the GNU Free Documentation License. You can get detailed informations about the authors and licences of pictures at the end of this book.

The content of this book was generated collaboratively by volunteers. Please be advised that nothing found here has necessarily been reviewed by people with the expertise required to provide you with complete, accurate or reliable information. Some information in this book maybe misleading or wrong. The Publisher does not guarantee the validity of the information found here. If you need specific advice (f.e. in fields of medical, legal, financial, or risk management questions) please contact a professional who is licensed or knowledgeable in that area.

Any brand names and product names mentioned in this book are subject to trademark, brand or patent protection and are trademarks or registered trademarks of their respective holders. The use of brand names, product names, common names, trade names, product descriptions etc. even without a particular marking in this works is in no way to be construed to mean that such names may be regarded as unrestricted in respect of trademark and brand protection legislation and could thus be used by anyone.

Cover image: www.ingimage.com
Concerning the licence of the cover image please contact ingimage.

Publisher:
Claud Press is a trademark of
International Book Market Service Ltd., 17 Rue Meldrum, Beau Bassin, 1713-01 Mauritius
Email: info@bookmarketservice.com
Website: www.bookmarketservice.com

Published in 2012

Printed in: U.S.A., U.K., Germany. This book was not produced in Mauritius.

ISBN: 978-620-1-90171-1

Contents

Articles

References

Nokia_C2-02

Nokia C2-02

Manufacturer	Nokia
Compatible networks	GSM 900/1800
Availability by country	Q3 2011
Form factor	slider
Dimensions	103 x 51.4 x 17 mm
Weight	115 g
Operating system	Series 40 6th edition feature pack 1
Memory	10 MB
Battery	BL-5C 3.7 V 1020 mAh
Display	2.6 inch resistive touch QVGA TFT (262 000 colors)
Rear camera	2 megapixels
Connectivity	Bluetooth 2.1+EDR, Micro-USB

The **Nokia C2-02** is a mobile telephone handset produced by Nokia. This is the first mobile handset released by Nokia that possesses a touchscreen in a "slider" phone form factor. Previously released touchscreen devices from Nokia using Series 40 Operating System have been in "candybar" form factor.

Features

The key feature of this phone is touch and type. It means that the phone has touch screen and alpha-numeric (12 key) keyboard but no navigation or softkeys. Other main features include: a 2.0 megapixel camera, Maps, Bluetooth 2.1 + EDR, Flash Lite 3.0 and MIDP Java 2.1 with additional Java APIs.

Specification sheet

Type	Specification
Modes	GSM 900 / GSM 1800
Regional Availability	Global
Weight	115 g
Dimensions	103 x 51.4 x 17 mm
Form Factor	Slider
Battery Life	Talk Time: 5 hours (GSM), Standby: 25 days (GSM)
Battery Type	BL-5C 3.7 V 1020 mAh
Display	Type: TFT Colors: 262 000 (18-bit) Size 2.4" Resolution: 240 x 320 pixels (QVGA)
Platform / OS	BB5 / Nokia Series 40, 6th Edition feature pack 1
Memory	10 MB
Digital TTY/TDD	Yes
Multiple Languages	Yes

Ringer Profiles	Yes
Vibrate	Yes
Bluetooth	Supported Profiles: DUN, FTP, GAP, GOEP, HFP, HSP, OPP, PAN, PBAP, SAP, SDAP, SPP
PC Sync	Yes
USB	Micro-USB
Multiple Numbers per Name	Yes
Voice Dialing	No
Custom Graphics	Yes
Custom Ringtones	Yes
Data-Capable	Yes
Flight Mode	Yes
Packet Data	Technology: GPRS, EDGE (EGPRS)
WLAN	No
WAP / Web Browser	HTML over TCP/IP, WAP 2.0, Proxy Browser with Compression Technology, XHTML over TCP/IP
Predictive Text Entry	T9
Side Keys	volume keys on right
Memory Card Slot	Card Type: microSD up to 32 GB.
Email Client	Protocols Supported: IMAP4, POP3, SMTP supports attachments
MMS	MMS 1.2 / SMIL
Text Messaging	2-Way: Yes
FM Radio	Stereo: Yes
Music Player	Supported Formats: AAC, AAC+, AMR-NB, AMR-WB, eAAC+, MIDI Tones (poly 64), Mobile XMF, MP3, MP4, NRT, True tones, WAV, WMA
Camera	Resolution: 2 megapixel (1600 x 1200)
Streaming Video	No
Video Capture	15 fps / 3GPP formats (H.263), H.264/AVC, MPEG-4, WMV
Alarm	Yes
Calculator	Yes
Calendar	Yes
SyncML	Yes
To-Do List	Yes
Voice Memo	Yes
Games	Yes
Java ME	Version: MIDP 2.1, CLDC 1.1 supported JSRs: 75, 82, 118, 135, 139, 172, 177, 179, 184, 205, 211, 226, 234, 248, Nokia UI API 1.1b (Includes Gesture API and Frame Animator API)
Headset Jack	Yes (3.5 mm)
Speaker Phone	Yes
Latest Firmware Version	v07.48

References

External links

- Nokia C2-02 Device specifiction at Forum Nokia (http://www.developer.nokia.com/Devices/Device_specifications/C2-02/)

Nokia

Nokia Corporation

NOKIA	
Type	Public
Traded as	OMX: NOK1V [1], NYSE: NOK [2], FWB: NOA3 [3]
Industry	Telecommunications equipment Internet Computer software
Founded	Tampere, Grand Duchy of Finland (1865) incorporated in Nokia (1871)
Founder(s)	Fredrik Idestam Leo Mechelin
Headquarters	Espoo, Finland
Area served	Worldwide
Key people	Risto Siilasmaa (Chairman) Stephen Elop (President & CEO)
Products	Mobile phones Smartphones Mobile computers Networks (See products listing)
Services	Maps and navigation, music, messaging and media Software solutions (See services listing)
Revenue	▼ €38.65 billion (2011)[4]
Operating income	▼ € -1.073 billion (2011)[4]
Net income	▼ € -1.164 billion (2011)[4]
Total assets	▼ €36.20 billion (2011)[4]
Total equity	▼ €11.87 billion (2011)[4]
Employees	122,148 (2012)[5]
Divisions	Mobile Solutions Mobile Phones Markets
Subsidiaries	Nokia Siemens Networks Navteq Vertu Qt Development Frameworks
Website	Nokia.com [6]

Nokia Corporation (Finnish pronunciation: [ˈnokiɑ], English /ˈnɒkiə/) (OMX: NOK1V [1], NYSE: NOK [2], FWB: NOA3 [3]) is a Finnish multinational communications and information technology corporation headquartered in Keilaniemi, Espoo, Finland.[7] Its principal products are mobile telephones and portable IT devices. It also offers Internet services including applications, games, music, maps, media and messaging through its Ovi platform, and free-of-charge digital map information and navigation services through its wholly owned subsidiary Navteq.[8] Nokia has a joint venture with Siemens, Nokia Siemens Networks, which provides telecommunications network equipment and services.[9]

Nokia has around 122,000 employees across 120 countries, sales in more than 150 countries and annual revenues of around €38 billion.[4] As of 2012 it is the world's second-largest mobile phone maker by unit sales (after Samsung), with a global market share of 22.5% in the first quarter.[10] Nokia is a public limited-liability company listed on the Helsinki, Frankfurt, and New York stock exchanges.[11] It is the world's 143rd-largest company measured by 2011 revenues according to the *Fortune Global 500*.[12]

Nokia was the world's largest vendor of mobile phones from 1998 to 2012.[10] However, over the past five years it has suffered declining market share as a result of the growing use of smartphones, principally the Apple iPhone and devices running on Google's Android operating system. As a result, its share price has fallen from a high of US$40 in 2007 to under US$3 in 2012.[13][14] Since February 2011, Nokia has had a strategic partnership with Microsoft, as part of which all Nokia smartphones will incorporate Microsoft's Windows Phone operating system (replacing Symbian). Nokia unveiled its first Windows Phone handsets, the Lumia 710 and 800, in October 2011.[15]

History

Pre-telecommunications era

Fredrik Idestam, co-founder of Nokia.

Statesman Leo Mechelin, co-founder of Nokia.

The predecessors of the modern Nokia were the Nokia Company (Nokia Aktiebolag), Finnish Rubber Works Ltd (Suomen Gummitehdas Oy) and Finnish Cable Works Ltd (Suomen Kaapelitehdas Oy).[16]

Nokia's history started in 1865 when mining engineer Fredrik Idestam established a groundwood pulp mill on the banks of the Tammerkoski rapids in the town of Tampere, in southwestern Finland in the Russian Empire and started manufacturing paper.[17] In 1868, Idestam built a second mill near the town of Nokia, fifteen kilometres (nine miles) west of Tampere by the Nokianvirta river, which had better resources for hydropower production.[18] In 1871, Idestam, with the help of his close friend statesman Leo Mechelin, renamed and transformed his firm into a share company, thereby founding the Nokia Company, the name it is still known by today.[18]

Toward the end of the 19th century, Mechelin's wishes to expand into the electricity business were at first thwarted by Idestam's opposition. However, Idestam's retirement from the management of the company in 1896 allowed Mechelin to become the company's chairman (from 1898 until 1914) and sell most shareholders on his plans, thus realizing his vision.[18] In 1902, Nokia added electricity generation to its business activities.[17]

Industrial conglomerate

In 1898, Eduard Polón founded Finnish Rubber Works, manufacturer of galoshes and other rubber products, which later became Nokia's rubber business.[16] At the beginning of the 20th century, Finnish Rubber Works established its factories near the town of Nokia and they began using Nokia as its product brand.[19] In 1912, Arvid Wickström founded Finnish Cable Works, producer of telephone, telegraph and electrical cables and the foundation of Nokia's cable and electronics businesses.[16] At the end of the 1910s, shortly after World War I, the Nokia Company was nearing bankruptcy.[20] To ensure the continuation of electricity supply from Nokia's generators, Finnish Rubber Works acquired the business of the insolvent company.[20] In 1922, Finnish Rubber Works acquired Finnish Cable Works.[21] In 1937, Verner Weckman, a sport wrestler and Finland's first Olympic Gold medalist, became president of Finnish Cable Works, after 16 years as its technical director.[22] After World War II, Finnish Cable Works supplied cables to the Soviet Union as part of Finland's war reparations. This gave the company a good foothold for later trade.[22]

The three companies, which had been jointly owned since 1922, were merged to form a new industrial conglomerate, Nokia Corporation in 1967 and paved the way for Nokia's future as a global corporation.[23] The new company was involved in many industries, producing at one time or another paper products, car and bicycle tires, footwear (including rubber boots), communications cables, televisions and other consumer electronics, personal computers, electricity generation machinery, robotics, capacitors, military communications and equipment (such as the SANLA M/90 device and the M61 gas mask for the Finnish Army), plastics, aluminium and chemicals.[24] Each business unit had its own director who reported to the first Nokia Corporation President, Björn Westerlund. As the president of the Finnish Cable Works, he had been responsible for setting up the company's first electronics department in 1960, sowing the seeds of Nokia's future in telecommunications.[25]

Eventually, the company decided to leave consumer electronics behind in the 1990s and focused solely on the fastest growing segments in telecommunications.[26] Nokian Tyres, manufacturer of tires, split from Nokia Corporation to form its own company in 1988[27] and two years later Nokian Footwear, manufacturer of rubber boots, was founded.[19] During the rest of the 1990s, Nokia divested itself of all of its non-telecommunications businesses.[26]

Telecommunications era

The seeds of the current incarnation of Nokia were planted with the founding of the electronics section of the cable division in 1960 and the production of its first electronic device in 1962: a pulse analyzer designed for use in nuclear power plants.[25] In the 1967 fusion, that section was separated into its own division, and began manufacturing telecommunications equipment. A key CEO and subsequent Chairman of the Board was *vuorineuvos* Björn "Nalle" Westerlund (1912–2009), who founded the electronics department and let it run at a loss for 15 years.

Networking equipment

In the 1970s, Nokia became more involved in the telecommunications industry by developing the Nokia DX 200, a digital switch for telephone exchanges. The DX 200 became the workhorse of the network equipment division. Its modular and flexible architecture enabled it to be developed into various switching products.[28] In 1984, development of a version of the exchange for the Nordic Mobile Telephony network was started.[29]

For a while in the 1970s, Nokia's network equipment production was separated into *Telefenno*, a company jointly owned by the parent corporation and by a company owned by the Finnish state. In 1987, the state sold its shares to Nokia and in 1992 the name was changed to Nokia Telecommunications.

In the 1970s and 1980s, Nokia developed the Sanomalaitejärjestelmä ("Message device system"), a digital, portable and encrypted text-based communications device for the Finnish Defence Forces.[30] The current main unit used by the Defence Forces is the Sanomalaite M/90 (SANLA M/90).[31]

First mobile phones

The technologies that preceded modern cellular mobile telephony systems were the various "0G" pre-cellular mobile radio telephony standards. Nokia had been producing commercial and some military mobile radio communications technology since the 1960s, although this part of the company was sold some time before the later company rationalization. Since 1964, Nokia had developed VHF radio simultaneously with Salora Oy. In 1966, Nokia and Salora started developing the ARP standard (which stands for Autoradiopuhelin, or *car radio phone* in English), a car-based mobile radio telephony system and the first commercially operated public mobile phone network in Finland. It went online in 1971 and offered 100% coverage in 1978.[34]

The Mobira Cityman 150, Nokia's NMT-900 mobile phone from 1989 (left), compared to the Nokia 1100 from 2003.[32] The Mobira Cityman line was launched in 1987.[33]

In 1979, the merger of Nokia and Salora resulted in the establishment of Mobira Oy. Mobira began developing mobile phones for the NMT (Nordic Mobile Telephony) network standard, the first-generation, first fully automatic cellular phone system that went online in 1981.[35] In 1982, Mobira introduced its first car phone, the Mobira Senator for NMT-450 networks.[35]

Nokia bought Salora Oy in 1984 and now owning 100% of the company, changed the company's telecommunications branch name to Nokia-Mobira Oy. The Mobira Talkman, launched in 1984, was one of the world's first transportable phones. In 1987, Nokia introduced one of the world's first handheld phones, the Mobira Cityman 900 for NMT-900 networks (which, compared to NMT-450, offered a better signal, yet a shorter roam). While the Mobira Senator of 1982 had weighed 9.8 kg (**unknown operator: u'strong'** lb) and the Talkman just under 5 kg (**unknown operator: u'strong'** lb), the Mobira Cityman weighed only 800 g (**unknown operator: u'strong'** oz) with the battery and had a price tag of 24,000 Finnish marks (approximately €4,560).[33] Despite the high price, the first phones were almost snatched from the sales assistants' hands. Initially, the mobile phone was a "yuppie" product and a status symbol.[24]

Nokia's mobile phones got a big publicity boost in 1987, when Soviet leader Mikhail Gorbachev was pictured using a Mobira Cityman to make a call from Helsinki to his communications minister in Moscow. This led to the phone's nickname of the "Gorba".[33]

In 1988, Jorma Nieminen, resigning from the post of CEO of the mobile phone unit, along with two other employees from the unit, started a notable mobile phone company of their own, Benefon Oy (since renamed to GeoSentric).[36] One year later, Nokia-Mobira Oy became Nokia Mobile Phones.

Involvement in GSM

Nokia was one of the key developers of GSM (Global System for Mobile Communications),[37] the second-generation mobile technology which could carry data as well as voice traffic. NMT (Nordic Mobile Telephony), the world's first mobile telephony standard that enabled international roaming, provided valuable experience for Nokia for its close participation in developing GSM, which was adopted in 1987 as the new European standard for digital mobile technology.[38][39]

Nokia delivered its first GSM network to the Finnish operator Radiolinja in 1989.[40] The world's first commercial GSM call was made on 1 July 1991 in Helsinki, Finland over a Nokia-supplied network, by then Prime Minister of Finland Harri Holkeri, using a prototype Nokia GSM phone.[40] In 1992, the first GSM phone, the Nokia 1011, was launched.[40][41] The model number refers to its launch date, 10 November.[41] The Nokia 1011 did not yet employ Nokia's characteristic ringtone, the Nokia tune. It was introduced as a ringtone in 1994 with the Nokia 2100 series.[42]

GSM's high-quality voice calls, easy international roaming and support for new services like text messaging (SMS) laid the foundations for a worldwide boom in mobile phone use.[40] GSM came to dominate the world of mobile telephony in the 1990s, in mid-2008 accounting for about three billion mobile telephone subscribers in the world, with more than 700 mobile operators across 218 countries and territories. New connections are added at the rate of 15 per second, or 1.3 million per day.[43]

Personal computers and IT equipment

In the 1980s, Nokia's computer division Nokia Data produced a series of personal computers called MikroMikko.[44] MikroMikko was Nokia Data's attempt to enter the business computer market. The first model in the line, MikroMikko 1, was released on 29 September 1981,[45] around the same time as the first IBM PC. However, the personal computer division was sold to the British ICL (International Computers Limited) in 1991, which later became part of Fujitsu.[46] MikroMikko remained a trademark of ICL and later Fujitsu. Internationally the MikroMikko line was marketed by Fujitsu as the ErgoPro.

The Nokia Booklet 3G mini laptop.

Fujitsu later transferred its personal computer operations to Fujitsu Siemens Computers, which shut down its only factory in Espoo, Finland (in the Kilo district, where computers had been produced since the 1960s) at the end of March 2000,[47][48] thus ending large-scale PC manufacturing in the country. Nokia was also known for producing very high quality CRT and early TFT LCD displays for PC and larger systems application. The Nokia Display Products' branded business was sold to ViewSonic in 2000.[49] In addition to personal computers and displays, Nokia used to manufacture DSL modems and digital set-top boxes.

Nokia re-entered the PC market in August 2009 with the introduction of the Nokia Booklet 3G mini laptop.[50]

Challenges of growth

In the 1980s, during the era of its CEO Kari Kairamo, Nokia expanded into new fields, mostly by acquisitions. In the late 1980s and early 1990s, the corporation ran into serious financial problems, a major reason being its heavy losses by the television manufacturing division and businesses that were just too diverse.[51] These problems, and a suspected total burnout, probably contributed to Kairamo taking his own life in 1988. After Kairamo's death, Simo Vuorilehto became Nokia's Chairman and CEO. In 1990–1993, Finland underwent severe economic depression,[52] which also struck Nokia. Under Vuorilehto's management, Nokia was severely overhauled. The company responded by streamlining its telecommunications divisions, and by divesting itself of the television and PC divisions.[53]

The Nokia House, Nokia's head office located by the Gulf of Finland in Keilaniemi, Espoo, was constructed between 1995 and 1997. It is the workplace of more than 1,000 Nokia employees.[24]

Probably the most important strategic change in Nokia's history was made in 1992, however, when the new CEO Jorma Ollila made a crucial strategic decision to concentrate solely on telecommunications.[26] Thus, during the rest of the 1990s, the rubber, cable and consumer electronics divisions were gradually sold as Nokia continued to divest itself of all of its non-telecommunications businesses.[26]

As late as 1991, more than a quarter of Nokia's turnover still came from sales in Finland. However, after the strategic change of 1992, Nokia saw a huge increase in sales to North America, South America and Asia.[54] The exploding worldwide popularity of mobile telephones, beyond even Nokia's most optimistic predictions, caused a logistics crisis in the mid-1990s.[55] This prompted Nokia to overhaul its entire logistics operation.[56] By 1998, Nokia's focus on telecommunications and its early investment in GSM technologies had made the company the world's largest mobile phone manufacturer,[54] a position it would hold for the next 14 consecutive years until 2012. Between 1996 and 2001, Nokia's turnover increased almost fivefold from 6.5 billion euros to 31 billion euros.[54] Logistics continues to be one of Nokia's major advantages over its rivals, along with greater economies of scale.[57][58]

Recent history

Product releases

Nokia launched its Nokia 1100 handset in 2003,[32] with over 200 million units shipped, was the best-selling mobile phone of all time and the world's top-selling consumer electronics product.[59] Also that year, Nokia is cited in Michael Saylor's 2012 book, *The Mobile Wave: How Mobile Intelligence Will Change Everything*, as one of the first players in the mobile space to recognize that there was a market opportunity in combining a game console and a mobile phone (both of which many gamers were carrying in 2003) into the N-Gage. The N-Gage was a mobile phone and game console meant to lure gamers away from the Game Boy Advance, though it cost twice as much and was said to resemble a taco.[60]

Reduction in size of Nokia mobile phones

In May 2007, Nokia released its first touch screen phone, the Nokia 7710, which was also a huge success. In November 2007, Nokia announced and released the Nokia N82, its first Nseries phone with Xenon flash. At the Nokia World conference in December 2007, Nokia announced their "Comes With Music" program: Nokia device buyers are to receive a year of complimentary access to music downloads.[61] The service became commercially available in the second half of 2008.

Evolution of the Nokia Communicator. Models 9000, 9110, 9210, 9300 and 9500 shown.

Nokia Productions was the first ever mobile filmmaking project directed by Spike Lee. Work began in April 2008, and the film premiered in October 2008.[62]

In 2008, Nokia released the Nokia E71 which was marketed to directly compete with the other BlackBerry-type devices offering a full "qwerty" keyboard and cheaper prices. Nokia announced in August 2009 that they will be selling a high-end Windows-based mini laptop called the Nokia Booklet 3G.[50] On 2 September 2009, Nokia launched two new music and social networking phones, the X6 and X3.[63] The Nokia X6 features 32GB of on-board memory with a 3.2" finger touch interface and comes with a music playback time of 35 hours. The Nokia X3 is a first series 40 Ovi Store-enabled device. The X3 is a music device that comes with stereo speakers, built-in FM radio, and a 3.2 megapixel camera. On 10 September 2009, Nokia unveiled the 7705 Twist, a phone sporting a square shape that swivels open to reveal a full QWERTY keypad, featuring a 3 megapixel camera, web browsing, voice commands and weighting around 3.44 ounces (**unknown operator: u'strong'** g).[64] On 9 August 2012, Nokia launched for the Indian market two new Asha range of handsets equipped with cloud accelerated Nokia browser, helping users browse the Internet faster and lower their spend on data charges. [65]

Symbian

Nokia N8 running on Symbian^3 OS was Nokia's first camera phone to feature a 12 megapixel camera with Carl Zeiss Optics

Originally Nokia phones had a custom Nokia OS operating system developed specifically for Nokia mobile phones.

The first Nseries device, the N90, utilised the older Symbian OS 8.1 mobile operating system, as did the N70. Subsequently Nokia switched to using SymbianOS 9 for all later Nseries devices (except the N72, which was based on the N70). Newer Nseries devices incorporate newer revisions of SymbianOS 9 that include Feature Packs. The N800, N810, N900, N9 and N950 are as of April 2012 the only Nseries devices (therefore excluding Lumia devices) to not use Symbian OS. They use the Linux-based Maemo.[66]

Nokia stated that Maemo would be developed alongside Symbian. Maemo had since (Maemo "6" and beyond) merged with Intel's Moblin, and became MeeGo. MeeGo was later canceled and a development is now continued under name Tizen.

The Nokia N8 is the first device to function on the Symbian^3 mobile operating system.

Nokia revealed that the N8 will be the last device in its flagship N-series devices to ship with Symbian OS.[67][68]

Instead, Nokia will use Microsoft Windows Phone for its high-end flagship Lumia devices, and revealed the Nokia N9 will function on the MeeGo mobile operating system.

Alliance with Microsoft

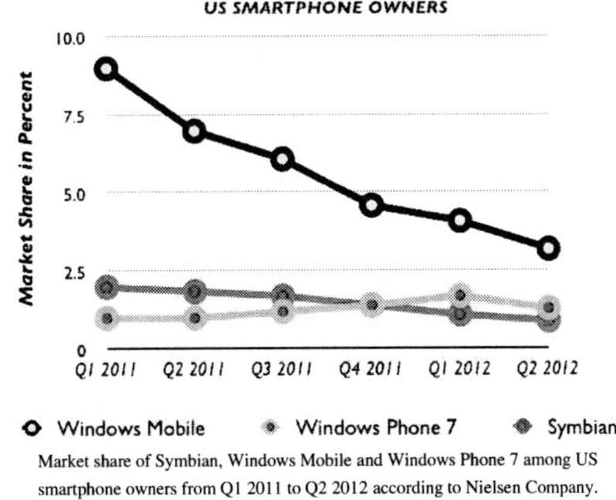

Market share of Symbian, Windows Mobile and Windows Phone 7 among US smartphone owners from Q1 2011 to Q2 2012 according to Nielsen Company.

On 11 February 2011, Nokia's CEO Stephen Elop, a former Microsoft employee, unveiled a new strategic alliance with Microsoft, and announced it would replace Symbian and MeeGo with Microsoft's Windows Phone operating system[69][70] except for mid-to-low-end devices, which would continue to run under Symbian. Nokia was also to invest into the Series 40 platform and release a single MeeGo product in 2011.[71]

As part of the restructuring plan, Nokia planned to reduce spending on research and development, instead customising and enhancing the software line for Windows Phone 7.[72] Nokia's "applications and content store" (Ovi) becomes integrated into the Windows Phone Marketplace, and Nokia Maps is at the heart of Microsoft's Bing and AdCenter. Microsoft provides developer tools to Nokia to replace the Qt framework, which is not supported by Windows Phone 7 devices.[73]

Symbian became described by Elop as a "franchise platform" with Nokia planning to sell 150 million Symbian devices after the alliance was set up. MeeGo emphasis was on longer-term exploration, with plans to ship "a MeeGo-related product" later in 2012. Microsoft's search engine, Bing was to become the search engine for all Nokia phones. Nokia also intended to get some level of customisation on WP7.[74]

After this announcement, Nokia's share price fell about 14%, its biggest drop since July 2009.[75]

As Nokia was the largest mobile phone manufacturer worldwide at the time,[76] it was suggested the alliance would make Microsoft's Windows Phone 7 a stronger contender against Android and iOS.[73] In June 2011 Nokia was overtaken by Apple as the world's biggest smartphone maker by volume.[77] In August 2011 Chris Weber, head of Nokia's subsidiary in the U.S., stated "*The reality is if we are not successful with Windows Phone, it doesn't matter what we do (elsewhere).*" He further added "*North America is a priority for Nokia (...) because it is a key market for Microsoft.*"[78]. Additionally, while announcing an alliance with Groupon, Elop declared "The competition... is not with other device manufacturers, it's with Google."[79]

European carriers have stated that Nokia Windows phones are not good enough to compete with Apple iPhone or Samsung Galaxy phones, that "they are overpriced for what is not an innovative product" and that "No one comes into the store and asks for a Windows phone".[80]

In June 2012, Nokia chairman Risto Siilasmaa told journalists that Nokia had a back-up plan in the eventuality that Windows Phone failed to be sufficiently successful in the market.[81][82]

Plant movements

Nokia opened its Komárom, Hungary mobile phone factory on 5 May 2000.[83]

In March 2007, Nokia signed a memorandum with Cluj County Council, Romania to open a new plant near the city in Jucu commune.[84][85][86] Moving the production from the Bochum, Germany factory to a low wage country created an uproar in Germany.[87][88] Nokia recently moved its North American Headquarters to Sunnyvale.

Reorganizations

In April 2003, the troubles of the networks equipment division caused the corporation to resort to similar streamlining practices on that side, including layoffs and organizational restructuring.[89] This diminished Nokia's public image in Finland,[90][91] and produced a number of court cases and an episode of a documentary television show critical of Nokia.[92]

On February 2006, Nokia and Sanyo announced a memorandum of understanding to create a joint venture addressing the CDMA handset business. But in June, they announced ending negotiations without agreement. Nokia also stated its decision to pull out of CDMA research and development, to continue CDMA business in selected markets.[93][94][95]

In June 2006, Jorma Ollila left his position as CEO to become the chairman of Royal Dutch Shell[96] and to give way for Olli-Pekka Kallasvuo.[97][98]

In May 2008, Nokia announced on their annual stockholder meeting that they want to shift to the Internet business as a whole. Nokia no longer wants to be seen as the telephone company. Google, Apple and Microsoft are not seen as natural competition for their new image but they are considered as major important players to deal with.[99]

In November 2008, Nokia announced it was ceasing mobile phone distribution in Japan.[100] Following early December, distribution of Nokia E71 is cancelled, both from NTT docomo and SoftBank Mobile. Nokia Japan retains global research & development programs, sourcing business, and an MVNO venture of Vertu luxury phones, using docomo's telecommunications network.

In February 2012, Nokia anonunced it was laying off 4000 employees to move manufacturing from Europe and Mexico to Asia.[101]

In March 2012, Nokia annonunced it was laying off 1000 employess from its Salo, Finland factory to focus on software.[102]

Acquisitions

For a more comprehensive list, see List of acquisitions by Nokia.

On 22 September 2003, Nokia acquired Sega.com, a branch of Sega which became the major basis to develop the Nokia N-Gage device.[103]

The Nokia E55 from the business segment of the Eseries range

On 16 November 2005, Nokia and Intellisync Corporation, a provider of data and PIM synchronization software, signed a definitive agreement for Nokia to acquire Intellisync.[104] Nokia completed the acquisition on 10 February 2006.[105]

On 19 June 2006, Nokia and Siemens AG announced the companies would merge their mobile and fixed-line phone network equipment businesses to create one of the world's largest network firms, Nokia Siemens Networks.[106] Each company has a 50% stake in the infrastructure company, and it is headquartered in Espoo, Finland. The companies predicted annual sales of €16 bn and cost savings of €1.5 bn a year by 2010. About 20,000 Nokia employees were transferred to this new company.

On 8 August 2006, Nokia and Loudeye Corp. announced that they had signed an agreement for Nokia to acquire online music distributor Loudeye Corporation for approximately US $60 million.[107] The company has been developing this into an online music service in the hope of using it to generate handset sales. The service, launched on 29 August 2007, is aimed to rival iTunes. Nokia completed the acquisition on 16 October 2006.[108]

In July 2007, Nokia acquired all assets of Twango, the comprehensive media sharing solution for organizing and sharing photos, videos and other personal media.[109][110]

In September 2007, Nokia announced its intention to acquire Enpocket, a supplier of mobile advertising technology and services.[111]

In October 2007, pending shareholder and regulatory approval, Nokia bought Navteq, a U.S.-based supplier of digital mapping data, for a price of $8.1 billion.[8][112] Nokia finalized the acquisition on 10 July 2008.[113]

In September, 2008, Nokia acquired OZ Communications, a privately held company with approximately 220 employees headquartered in Montreal, Canada.[114]

On 24 July 2009, Nokia announced that it will acquire certain assets of cellity, a privately owned mobile software company which employs 14 people in Hamburg, Germany.[115] The acquisition of cellity was completed on 5 August 2009.[116]

On 11 September 2009, Nokia announced the acquisition of "certain assets of Plum Ventures, Inc, a privately held company which employed approximately 10 people with main offices in Boston, Massachusetts. Plum will complement Nokia's Social Location services".[117]

On 28 March 2010, Nokia announced the acquisition of Novarra, the mobile web browser firm from Chicago. Terms of the deal were not disclosed. Novarra is a privately held company based in Chicago, IL and provider of a mobile browser and service platform and has more than 100 employees.[118]

On 10 April 2010, Nokia announced its acquisition of MetaCarta, whose technology was planned to be used in the area of local search, particularly involving location and other services. Financial details of acquisition were not disclosed.[119]

Nokia has acquired Smarterphone in 2012.[120] Also Nokia acquired Scalado in 2012.[121]

Financial difficulties and restructuring

Amid falling sales, Nokia posted a loss of 368 million euros for Q2 2011, while in Q2 2010 had still a profit of 227 million euros. On September 2011, Nokia has announced it will lose another 3,500 jobs worldwide, including the closure of its Cluj factory in Romania.[122]

On 8 February 2012 Nokia Corp. said to cut around 4,000 jobs at smartphone manufacturing plants in Europe by the end of 2012 to move assembly closer to component supplier in Asia. It plans to cut 2,300 of the 4,400 jobs in Hungary, 700 out of 1,000 jobs in Mexico, and 1,000 out of 1,700 factory jobs in Finland.[123]

On 14 June 2012, Nokia announced to cut 10,000 jobs globally by the end of 2013[124] and shut production and research sites in Finland, Germany and Canada inline with continues loss and the stock fell to the lowest since 1996. Today, Nokia's market value is below $10 billion.[125]

In total, according to actualized and planned laid-offs Nokia will have laid off 24,500 employees by the end of 2013. Nokia has already laid off 7,000 employees in the first stage: 4,000 staff and transferred also 3,000 to services firm Accenture. Nokia also closed its factory in Cluj, Romania that decreased the workforce by 2,000 employees, and restructured the Location & Commerce business unit that decreased the workforce by 1,200 employees. In February 2012, Nokia unveiled a plan to cut 4,000 more jobs at its plants in Finland, Hungary and Mexico as it moves smartphone assembly work to Asia. The most recent plan is to cut further 10,000 jobs globally by the end of 2013.[126] Nokia had 66,267 personnel in its Devices&Services, NAVTEQ and Corporate Common Functions units combined, this has been calculated by subtracting the personnel of Nokia Siemens Networks from the total personnel of Nokia Group based on the full year report of 2010.[127] Therefore, the personnel would decrease by approximately 36 percent by the end of 2013 when compared to the end of 2010 that best depicts the lay-offs that have resulted from the strategy change in February 2011 and competition in the central mobile phone business units recently.

On 18 June 2012 Moody's downgraded Nokia rating to junk.[128] Nokia CEO admitted on 28 June 2012 that company's inability to foresee rapid changes in mobile phone industry was one of the major reasons for the problems company was facing.[129]

Corporate affairs

Corporate structure

Divisions

Since 1 July 2010, Nokia comprises three business groups: **Mobile Solutions**, **Mobile Phones** and **Markets**.[130] The three units receive operational support from the **Corporate Development Office**, led by Kai Öistämö, which is also responsible for exploring corporate strategic and future growth opportunities.[130]

On 1 April 2007, Nokia's Networks business group was combined with Siemens's carrier-related operations for fixed and mobile networks to form Nokia Siemens Networks, jointly owned by Nokia and Siemens and consolidated by Nokia.[131]

Mobile Solutions

Mobile Solutions is responsible for Nokia's portfolio of smartphones and mobile computers, including the more expensive multimedia and enterprise-class devices. The team is also responsible for a suite of internet services under the Ovi brand, with a strong focus on maps and navigation, music, messaging and media.[130] This unit is led by Anssi Vanjoki, along with Tero Ojanperä (for Services) and Alberto Torres (for MeeGo Computers).[130]

The Nokia N900, a Maemo 5 Linux based mobile Internet device and touchscreen smartphone from Nokia's Nseries portfolio.

Alberto Torres has stepped down.

Mobile Phones

Mobile Phones is responsible for Nokia's portfolio of affordable mobile phones, as well as a range of services that people can access with them, headed by Mary T. McDowell.[130] This unit provides the general public with mobile voice and data products across a range of devices, including high-volume, consumer oriented mobile phones. The devices are based on GSM/EDGE, 3G/W-CDMA and CDMA cellular technologies.

The Nokia E90, a Symbian smartphone from Nokia's Eseries portfolio.

At the end of the year 2007, Nokia managed to sell almost 440 million mobile phones which accounted for 40% of all global mobile phones sales.[132] In 2011, Nokia's market share in the mobile phone market had dropped to 27% (417 million phones).[133]

Anssi Vanjoki resigned a few days before Nokia World 2010 and under new leadership team Jo Harlow will look into the affairs of Smartphones portfolio.

On 27 April 2011, The Register reported that Nokia was secretly developing a new operating system called Meltemi aiming at the low-end market. It was believed it would be replacing the S30 and S40 operating systems. Due to low-end market customers' demand of having smartphone features in their feature phone, the OS would have included some features exclusive to high-end smartphones. On 26 July 2012, it was announced that Nokia had abandoned the Meltemi project as a cost-cutting measure.

Markets

Markets is responsible for Nokia's supply chains, sales channels, brand and marketing functions of the company, and is responsible for delivering mobile solutions and mobile phones to the market. The unit is headed by Niklas Savander.[130]

Subsidiaries

Nokia has several subsidiaries, of which the two most significant as of 2009 are Nokia Siemens Networks and Navteq.[130] Other notable subsidiaries include, but are not limited to Vertu, a British-based manufacturer and retailer of luxury mobile phones; Qt Software, a Norwegian-based software company, and OZ Communications, a consumer e-mail and instant messaging provider.

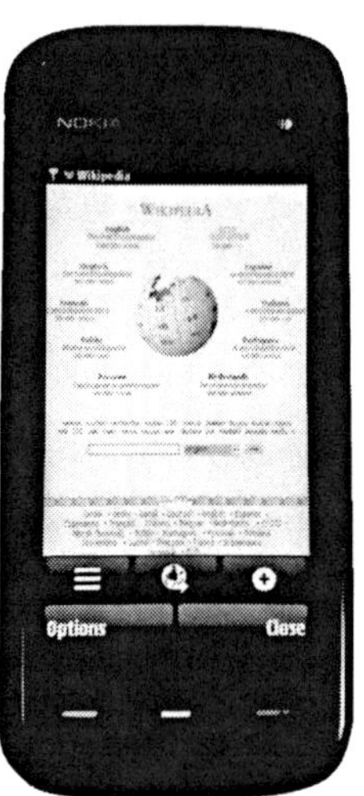

The Nokia 5800 XpressMusic, a touchscreen smartphone and portable entertainment device which emphasizes music and multimedia playback.

Until 2008 Nokia was the major shareholder in Symbian Limited, a software development and licensing company that produced Symbian OS, a smartphone operating system used by Nokia and other manufacturers. In 2008 Nokia acquired Symbian Ltd and, along with a number of other companies, created the Symbian Foundation to distribute the Symbian platform royalty free and as open source.

Nokia Siemens Networks

Nokia Siemens Networks (previously Nokia Networks) provides wireless and fixed network infrastructure, communications and networks service platforms, as well as professional services to operators and service providers.[130] Nokia Siemens Networks focuses in GSM, EDGE, 3G/W-CDMA and WiMAX radio access networks; core networks with increasing IP and multiaccess capabilities; and services.

On 19 June 2006 Nokia and Siemens AG announced the companies are to merge their mobile and fixed-line phone network equipment businesses to create one of the world's largest network firms, called Nokia Siemens Networks.[106] The Nokia Siemens Networks brand identity was subsequently launched at the 3GSM World Congress in Barcelona in February 2007.[134][135]

Navteq

Navteq is a Chicago, Illinois-based provider of digital map data and location-based content and services for automotive navigation systems, mobile navigation devices, Internet-based mapping applications, and government and business solutions.[130] Navteq was acquired by Nokia on 1 October 2007.[8] Navteq's map data is part of the Nokia Maps online service where users can download maps, use voice-guided navigation and other context-aware web services.[130] Nokia Maps is part of the Ovi brand of Nokia's Internet based online services.

Corporate governance

The control and management of Nokia is divided among the shareholders at a general meeting and the Nokia Leadership Team (left),[136] under the direction of the Board of Directors (right).[137] The Chairman and the rest of the Nokia Leadership Team members are appointed by the Board of Directors. Only the Chairman of the Nokia Leadership Team can belong to both, the Board of Directors and the Nokia Leadership Team. The Board of Directors' committees consist of the Audit Committee,[138] the Personnel Committee[139] and the Corporate Governance and Nomination Committee.[140][141]

The operations of the company are managed within the framework set by the Finnish Companies Act,[142] Nokia's Articles of Association[143] and Corporate Governance Guidelines,[144] and related Board of Directors adopted charters.

Nokia Leadership Team (as of April 2012) [136]
Stephen Elop (Chairman), b. 1963 President, CEO and Nokia Leadership Team Chairman of Nokia Corporation since 21 September 2010 Joined Nokia on 21 September 2010, Nokia Board member since 3 May 2011
Esko Aho, b. 1954 Executive Vice President, Corporate Relations and Responsibility Joined Nokia 2008, Nokia Leadership Team member since 2009 Former Prime Minister of Finland (1991–1995)
Marko Ahtisaari, b. 1969 Executive Vice President, Design Joined Nokia 2009, Nokia Leadership Team member since 1 February 2012
Jerri DeVard, b. 1958 Executive Vice President, Chief Marketing Officer Joined Nokia 2011, Nokia Leadership Team member since 1 January 2011
Colin Giles, b. 1963 Executive Vice President, Sales Joined Nokia 1992, Nokia Leadership Team member since 11 February 2011
Michael Halbherr, b. 1964 Executive Vice President, Location & Commerce Joined Nokia 2006, Nokia Leadership Team member since 1 July 2011
Jo Harlow, b. 1962 Executive Vice President, Smart Devices Joined Nokia 2003, Nokia Leadership Team member since 11 February 2011
Timo Ihamuotila, b. 1966 Executive Vice President, Chief Financial Officer With Nokia 1993–1996, rejoined 1999, Nokia Leadership Team member since 2007
Mary T. McDowell, b. 1964 Executive Vice President, Mobile Phones Joined Nokia 2004, Nokia Leadership Team member since 2004
Louise Pentland, b. 1972 Executive Vice President, Chief Legal Officer Joined Nokia 1998, Nokia Leadership Team member since 11 February 2011
Niklas Savander, b. 1962 Executive Vice President, Markets Joined Nokia 1997, Nokia Leadership Team member since 2006
Henry Tirri, b. 1956 Executive Vice President, Chief Technology Officer Joined Nokia 2004, Nokia Leadership Team member since 22 September 2011
Juha Äkräs, b. 1965 Executive Vice President, Human Resources Joined Nokia 1993, Nokia Leadership Team member since 2010
Dr. Kai Öistämö, b. 1964 Executive Vice President, Chief Development Officer Joined Nokia 1991, Nokia Leadership Team member since 2005

Board of Directors [137]
Risto Siilasmaa (Chairman), b. 1966 Board member since 2008, Chairman of the Board of Directors since 3 May 2012 Chairman of the Corporate Governance and Nomination Committee Founder and Chairman of F-Secure Corporation
Dame Marjorie Scardino (Vice Chairman), b. 1947 Board member since 2001, Vice Chairman since 2007 Member of the Corporate Governance and Nomination Committee, Member of the Personnel Committee Chief Executive Officer and member of the Board of Directors of Pearson PLC
Bruce Brown, b. 1958 Board member since 3 May 2012, Member of the Personnel Committee Chief Technology Officer of Procter & Gamble
Stephen Elop, b. 1963 Board member since 3 May 2011 President and CEO of Nokia Corporation, Chairman of the Nokia Nokia Leadership Team
Dr. Henning Kagermann, b. 1947 Board member since 2007, Chairman of the Personnel Committee, Member of the Corporate Governance and Nomination Committee Former CEO and Chairman of the Executive Board of SAP AG
Jouko Karvinen, b. 1957 Board member since 3 May 2011, Chairman of the Audit Committee, Member of the Corporate Governance and Nomination Committee CEO of Stora Enso Oyj
Helge Lund, b. 1962 Board member since 3 May 2011, Member of the Personnel Committee President and CEO of Statoil ASA
Isabel Marey-Semper, b. 1967 Board member since 2009, Member of the Audit Committee Director of Advanced Research of L'Oréal Group
Mårten Mickos, b. 1962 Board member since 3 May 2012 Chief Executive Officer of Eucalyptus Systems, Inc.
Elizabeth Nelson, b. 1960 Board member since 3 May 2012, Member of the Audit Committee Independent Corporate Advisor
Kari Stadigh, b. 1955 Board member since 3 May 2011, Member of the Personnel Committee Group CEO and President of Sampo plc

Former corporate officers

Chief Executive Officers		Chairmen of the Board of Directors [145]			
Björn Westerlund	1967–1977	Lauri J. Kivekäs	1967–1977	Simo Vuorilehto	1988–1990
Kari Kairamo	1977–1988	Björn Westerlund	1977–1979	Mika Tiivola	1990–1992
Simo Vuorilehto	1988–1992	Mika Tiivola	1979–1986	Casimir Ehrnrooth	1992–1999
Jorma Ollila	1992–2006	Kari Kairamo	1986–1988	Jorma Ollila	1999–2012
Olli-Pekka Kallasvuo	2006–2010				

International presence

In 2011 Nokia had 130,000 employees in 120 countries, sales in more than 150 countries, global annual revenue of over €38 billion, and operating loss of €1 billion.[4] It was the world's largest manufacturer of mobile phones in 2011, with global device market share of 23% in the second quarter.[76]

The Nokia Research Center, founded in 1986, is Nokia's industrial research unit consisting of about 500 researchers, engineers and scientists;[146][147] it has sites in seven countries: Finland, China, India, Kenya, Switzerland, the United Kingdom and the United States.[148] Besides its research centers, in 2001 Nokia founded (and owns) INdT – Nokia Institute of Technology, a R&D institute located in Brazil.[149] Nokia operates a total of 9 manufacturing facilities[11] located at Salo, Finland; Manaus, Brazil; Cluj, Romania; Beijing and Dongguan, China; Komárom, Hungary; Chennai, India; Reynosa, Mexico; and Changwon, South Korea.[84][150] Nokia's factory in Cluj was seized by the Romanian government in November 2011 to prevent a sale of the assets, after Nokia had accumulated a tax liability of US$ 10 million.[151] Nokia's industrial design department is headquartered in Soho in London, UK with significant satellite offices in Helsinki, Finland and Calabasas, California in the US.

Nokia is a public limited-liability company listed on the Helsinki, Frankfurt, and New York stock exchanges.[11] Nokia plays a very large role in the economy of Finland.[152][153] It is an important employer in Finland and several small companies have grown into large ones as its partners and subcontractors.[154] In 2009 Nokia contributed 1.6% to Finland's GDP, and accounted for about 16% of Finland's exports in 2006.[155]

In February 2012 Nokia announced that it was cutting 4,000 factory jobs in Finland, Hungary and Mexico (more than half of the 7,100 jobs at the three factories affected) and moving smartphone assembly to existing facilities in South Korea and China.[156]

Logos

Past

NOKIA

CONNECTING PEOPLE

Nokia introduced its "Connecting People" advertising slogan, coined by Ove Strandberg "HS Archives" (in Finnish). Helsingin Sanomat. 1 June 2003. . Retrieved 14 May 2008. and used since 1992. "NOKIA | Connecting Pople 1992 Vector Logo (AI EPS)". HDicon.com. . Retrieved 17 October 2010.This earlier version of the slogan used Times RomanTimes Roman SC (Small Caps) font.Pitkänen, Juhani (Nokia's Art Director) (3 September 2007). "Nokia Strategic Marketing, Brand Identity". Nokia Corporation. . Retrieved 14 May 2008.

Present

Nokia's current logo used since 2006, "NOKIA | Connecting Pople new Vector Logo (AI EPS)". HDicon.com. . Retrieved 17 October 2010. with the redesigned "Connecting People" slogan.This slogan originally used Nokia's proprietary 'Nokia Sans' font, designed by Erik Spiekermann. "Erik Spiekermann – Furniture, Designs & Home Decor". Design Within Reach. . Retrieved 7 January 2010. This was replaced in 2011 with the 'Nokia Pure' font designed by Dalton Maag. "Our New Typeface". Nokia Little Blog of Branding. . Retrieved 3 April 2012.

Stock

Nokia, a public limited liability company, is the oldest company listed under the same name on the Helsinki Stock Exchange (since 1915).[24] Nokia's shares are also listed on the Frankfurt Stock Exchange (since 1988) and New York Stock Exchange (since 1994).[11][24]

In 2007, Nokia was valued at €110 billion; as of May 2012, it was valued at €14.8 billion. [163]

For fiscal Q2 2011 ending in June 2011, Nokia reported a net loss of €492 million, despite a €430 million payment from Apple. Nokia cited decline in its mobile phone business as the primary cause of the loss.[164]

In Q1 2012 results were bleak. Nokia lost €1.34 billion. Revenue is down almost a third from a year ago.[165] By May 2012, Nokia share price had fallen 37.5 percent since the beginning of the year, and was down 61 percent in the last year.[166][167]

Corporate culture

Nokia's official corporate culture manifesto, *The Nokia Way*, emphasises the speed and flexibility of decision-making in a flat, networked organization, although the corporation's size necessarily imposes a certain amount of bureaucracy.[168]

The Nokia House, Nokia's head office in Keilaniemi, Espoo, Finland.

The official business language of Nokia is English. All documentation is written in English, and is used in official intra-company spoken communication and e-mail.

Until May 2007, the *Nokia Values* were Customer Satisfaction, Respect, Achievement, and Renewal. In May 2007, Nokia redefined its values after initiating a series of discussions worldwide as to what the new values of the company should be. Based on the employee suggestions, the new values were defined as: Engaging You, Achieving Together, Passion for Innovation and Very Human.[168]

Online services

.mobi and the Mobile Web

Nokia was the first proponent of a Top Level Domain (TLD) specifically for the Mobile Web and, as a result, was instrumental in the launch of the .mobi domain name extension in September 2006 as an official backer.[169][170] Since then, Nokia has launched the largest mobile portal, Nokia.mobi, which receives over 100 million visits a month.[171] It followed that with the launch of a mobile Ad Service to cater to the growing demand for mobile advertisement.[172]

Ovi

Ovi, announced on 29 August 2007, is the name for Nokia's "umbrella concept" Internet services.[173] Centered on Ovi.com, it is marketed as a "personal dashboard" where users can share photos with friends, download music, maps and games directly to their phones and access third-party services like Yahoo's Flickr photo site. It has some significance in that Nokia is moving deeper into the world of Internet services, where head-on competition with Microsoft, Google and Apple is inevitable.[174]

The services offered through Ovi include the Ovi Store (Nokia's application store), the Nokia Music Store, Nokia Maps, Ovi Mail, the N-Gage mobile gaming platform available for several S60 smartphones, Ovi Share, Ovi Files, and Contacts and Calendar.[175] The Ovi Store, the Ovi application store was launched in May 2009.[176] Prior to opening the Ovi Store, Nokia integrated its software Download! store, the stripped-down MOSH repository and the widget service WidSets into it.[177]

On 23 March 2010, Nokia announced launch of its online magazine called the *Nokia Ovi*. The 44-page magazine contains articles on products by Nokia, what Ovi stands for, tips and tricks on the usage of Nokia mini laptop Booklet 3G, latest reviews of mobile applications, news about the mobile maker's services and apps such as Ovi maps, files and mail. Users can download the magazine as a PDF or view it online from the Nokia website.[178]

My Nokia

Nokia offers a free personalised service to Nokia owners called My Nokia (located at my.nokia.com).[179] Registered My Nokia users can get free services as follows:

- Tips & tricks alerts through web, e-mail and also mobile text message.
- My Nokia Backup: A free online backup service for mobile contacts, calendar logs and also various other files. This service needs GPRS connection.
- Ringtones, wallpapers, screensavers, games and other things can be downloaded free of cost.

Comes With Music

In 2007 Nokia set up their "Nokia Comes With Music" service, in partnership with Universal Music Group International, Sony BMG, Warner Music Group, EMI, and hundreds of independent labels and music aggregators, to allow 12, 18, or 24 months of unlimited free-of-charge music downloads with the purchase of a Nokia Comes With Music edition phone. Files could be downloaded on mobile devices or personal computers, and kept permanently.[61]

In January 2011 Nokia withdrew this program in 27 countries, due to its failure to gain traction with customers or mobile network operators; existing subscribers could continue to download until their contracts ended. The service continued to be offered in China, India, Indonesia, Brazil, Turkey and South Africa where take-up had been better.[180]

Nokia Messaging

On 13 August 2008 Nokia launched a beta release of "Nokia Email service", a push e-mail service, since incorporated into Nokia Messaging.[181]

Nokia Messaging operates as a centralised, hosted service that acts as a proxy between the Nokia Messaging client and the user's e-mail server. The phone does not connect directly to the e-mail server, but instead sends e-mail credentials to Nokia's servers.[182] IMAP is used as the protocol to transfer emails between the client and the server.

Controversies

NSN's provision of intercept capability to Iran

In 2008, Nokia Siemens Networks, a joint venture between Nokia and Siemens AG, reportedly provided Iran's monopoly telecom company with technology that allowed it to intercept the Internet communications of its citizens to an unprecedented degree.[183] The technology reportedly allowed it to use deep packet inspection to read and even change the content of everything from "e-mails and Internet phone calls to images and messages on social-networking sites such as Facebook and Twitter". The technology "enables authorities to not only block communication but to monitor it to gather information about individuals, as well as alter it for disinformation purposes," expert insiders told *The Wall Street Journal*. During the post-election protests in Iran in June 2009, Iran's Internet access was reported to have slowed to less than a tenth of its normal speeds, and experts suspected this was due to the use of the interception technology.[184]

The joint venture company, Nokia Siemens Networks, asserted in a press release that it provided Iran only with a 'lawful intercept capability' "solely for monitoring of local voice calls". "Nokia Siemens Networks has not provided any deep packet inspection, web censorship or Internet filtering capability to Iran," it said.[185]

In July 2009, Nokia began to experience a boycott of their products and services in Iran. The boycott was led by consumers sympathetic to the post-election protest movement and targeted at those companies deemed to be collaborating with the Islamic regime. Demand for handsets fell and users began shunning SMS messaging.[186]

Lex Nokia

In 2009, Nokia heavily supported the passing of a law in Finland that allows companies to monitor their employees' electronic communications in cases of suspected information leaking.[187] Contrary to rumors, Nokia denied that the company would have considered moving its head office out of Finland if laws on electronic surveillance were not changed.[188] The law was enacted, but with strict requirements for implementation of its provisions. As of 2010, the law has become a dead letter; no corporation has implemented it. The Finnish media dubbed the name *Lex Nokia* for this law, named after the Finnish copyright law (the so-called *Lex Karpela*) a few years back.

Nokia–Apple patent dispute

In October 2009, Nokia filed a lawsuit against Apple Inc. in the U.S. District Court of Delaware citing Apple infringed on 10 of its patents related to wireless communication including data transfer.[189] Apple was quick to respond with a countersuit filed in December 2009 accusing Nokia of 11 patent infringements. Apple's General Counsel, Bruce Sewell went a step further by stating, "Other companies must compete with us by inventing their own technologies, not just by stealing ours." This resulted in an ugly spat between the two telecom majors with Nokia filing another suit, this time with the U.S. International Trade Commission (ITC), alleging Apple of infringing its patents in "virtually all of its mobile phones, portable music players, and computers."[190] Nokia went on to ask the court to bar all U.S. imports of the Apple products including the iPhone, Mac and the iPod. Apple countersued by filing a complaint with the ITC in January 2010, the details of which are yet to be confirmed.[189]

In June 2011, Apple settled with Nokia and agreed to an estimated one time payment of $600 million and royalties to Nokia.[191] The two companies also agreed on a cross-licensing patents for some of their patented technologies.[192][193]

Environmental record

Electronic products such as cell phones impact the environment both during production and after their useful life when they are discarded and turned into electronic waste. Nokia is listed in Greenpeace's Guide to Greener Electronics that scores leading electronics manufacturers according to their policies on sustainability, climate and energy and how green their products are. In November 2011 Nokia ranked 3rd out of 15 listed electronics companies, falling two places due to its weaker performance on the Energy criteria and scoring 4.9/10.[194]

All of Nokia's mobile phones are free of toxic polyvinyl chloride (PVC) since the end of 2005 and all new models of mobile phones and accessories launched in 2010 are on track to be free of brominated compounds, chlorinated flame retardants and antimony trioxide.[194]

Nokia's voluntary take-back programme to recycle old mobile phones spans 84 countries with almost 5,000 collection points.[195] However, the recycling rate of Nokia phones was only 3–5% in 2008, according to a global consumer survey released by Nokia.[196] The majority of old mobile phones are simply lying in drawers at home and very few old devices, about 4%, are being thrown into landfill and not recycled.[196]

All of Nokia's new models of chargers meet or exceed the Energy Star requirements.[197] Nokia aims to reduce its carbon dioxide emissions by at least 18 percent in 2010 from a baseline year of 2006 and cover 50 percent of its energy needs through renewable energy sources.[198] Greenpeace is challenging the company to use its influence at the political level as number 85 on the Fortune 500 to advocate for climate legislation and call for global greenhouse gas emissions to peak by 2015.[199]

Nokia is researching the use of recycled plastics in its products, which are currently used only in packaging but not yet in mobile phones.[200]

Since 2001, Nokia has provided eco declarations of all its products and since May 2010 provides Eco profiles for all its new products.[201] In an effort to further reduce their environmental impact in the future, Nokia released a new phone concept, Remade, in February 2008.[202] The phone has been constructed of solely recyclable materials.[202] The outer part of the phone is made from recycled materials such as aluminium cans, plastic bottles, and used car tires.[203] The screen is constructed of recycled glass, and the hinges have been created from rubber tires. The interior of the phone is entirely constructed with refurbished phone parts, and there is a feature that encourages energy saving habits by reducing the backlight to the ideal level, which then allows the battery to last longer without frequent charges.

Research cooperation with universities

Nokia is actively exploring and engaging in open innovation through selective research collaborations with major universities and institutions by sharing resources and leveraging ideas. Major research collaboration is with Tampere University of Technology based in Finland. Current collaborations include:[204]

- Aalto University School of Science and Technology, Finland
- École Polytechnique Fédérale de Lausanne, Switzerland
- ETH Zurich, Switzerland
- Massachusetts Institute of Technology, United States
- Stanford University, United States
- Tampere University of Technology, Finland
- Tsinghua University, China
- University of California, Berkeley, United States
- University of Cambridge, United Kingdom
- University of Southern California, United States

Awards and recognition

The Brand Trust Report [205] published by Trust Research Advisory has ranked Nokia in the 1st position among the brands in India.

References

[1] http://www.nasdaqomxnordic.com/shares/shareinformation?Instrument=HEX24311
[2] http://www.nyse.com/about/listed/quickquote.html?ticker=nok
[3] http://www.boerse-frankfurt.de/en/equities/search/result?name_isin_wkn=NOA3
[4] "Annual Results 2011" (http://www.results.nokia.com/results/Nokia_results2011Q4e.pdf) (PDF). Nokia Corporation. 26 January 2012. . Retrieved 2 March 2012.
[5] www.results.nokia.com/results/Nokia_results2012Q1e.pdf
[6] http://www.nokia.com/
[7] "Nokia in brief (2007)" (http://www.nokia.com/NOKIA_COM_1/About_Nokia/Sidebars_new_concept/Nokia_in_brief/InBriefJuly08.pdf) (PDF). Nokia Corporation. March 2008. . Retrieved 14 May 2008.
[8] "Nokia to acquire NAVTEQ" (http://www.nokia.com/A4136001?newsid=1157198) (Press release). Nokia Corporation. 1 October 2007. . Retrieved 22 March 2009.
[9] "Company" (http://www.nokiasiemensnetworks.com/global/AboutUs/Company/?languagecode=en). Nokia Siemens Networks. . Retrieved 14 July 2009.
[10] "Samsung overtakes Nokia in mobile phone shipments" (http://www.bbc.co.uk/news/business-17865117). BBC News. 27 April 2012. . Retrieved 27 May 2012.
[11] "Nokia – FAQ" (http://www.nokia.com/about-nokia/company/faq). Nokia Corporation. . Retrieved 16 March 2009.
[12] "*Global 500* 2011" (http://money.cnn.com/magazines/fortune/global500/2011/full_list/101_200.html). Fortune. 2011. . Retrieved 12 November 2011.

[13] https://www.google.com/finance?client=ob&q=NYSE:NOK

[14] http://247wallst.com/2012/05/14/almost-nothing-can-save-nokia-except-maybe-widespread-poverty/#ixzz1v8jU0K5J

[15] "Nokia's first Windows Phone 7 handset: Is it enough?" (http://www.bbc.co.uk/news/technology-15460569). BBC News. 26 October 2011. . Retrieved 27 October 2011.

[16] "Nokia – Nokia's first century – Story of Nokia" (http://www.nokia.com/about-nokia/company/story-of-nokia/nokias-first-century). Nokia Corporation. . Retrieved 16 March 2009.

[17] "Nokia – The birth of Nokia – Nokia's first century – Story of Nokia" (http://www.nokia.com/about-nokia/company/story-of-nokia/nokias-first-century/the-birth-of-nokia). Nokia Corporation. . Retrieved 16 March 2009.

[18] Helen, Tapio. "Idestam, Fredrik (1838–1916)" (http://www.kansallisbiografia.fi/english/?id=4296). Biographical Centre of the Finnish Literature Society. . Retrieved 22 March 2009.

[19] "Nokian Footwear: History" (http://www.nokianfootwear.fi/eng/our_story/). Nokian Footwear. . Retrieved 21 March 2009.

[20] Palo-oja, Ritva; Willberg, Leena (1998) (in Finnish). *Kumi – Kumin ja Suomen kumiteollisuuden historia*. Tampere, Finland: Tampere Museums. pp. 43–53. ISBN 978-951-609-065-1.

[21] "Finnish Cable Factory – Brief History" (http://web.archive.org/web/20070705233815/http://www.kaapelitehdas.fi/php/image.php?id=4856) (PDF). Kaapelitehdas.fi (http://www.kaapelitehdas.fi). Archived from the original (http://www.kaapelitehdas.fi/php/image.php?id=4856) on 5 July 2007. . Retrieved 16 March 2009.

[22] "Nokia – Verner Weckman – Nokia's first century – Story of Nokia" (http://www.nokia.com/about-nokia/company/story-of-nokia/nokias-first-century/verner-weckman). Nokia Corporation. . Retrieved 20 March 2009.

[23] "Nokia – The merger – Nokia's first century – Story of Nokia" (http://www.nokia.com/about-nokia/company/story-of-nokia/nokias-first-century/the-merger). Nokia Corporation. . Retrieved 16 March 2009.

[24] "Nokia – Towards Telecommunications" (http://www.nokia.com/NOKIA_COM_1/About_Nokia/Sidebars_new_concept/Broschures/TowardsTelecomms.pdf) (PDF). Nokia Corporation. August 2000. . Retrieved 5 June 2008.

[25] "Nokia – First electronic dept – Nokia's first century – Story of Nokia" (http://www.nokia.com/about-nokia/company/story-of-nokia/nokias-first-century/first-electronic-dept). Nokia Corporation. . Retrieved 16 March 2009.

[26] "Nokia – Jorma Ollila – Mobile revolution – Story of Nokia" (http://www.nokia.com/about-nokia/company/story-of-nokia/mobile-revolution/jorma-ollila). Nokia Corporation. . Retrieved 21 March 2009.

[27] "History in brief" (http://www.nokiantyres.com/history-in-brief). Nokian Tyres. . Retrieved 22 March 2009.

[28] Kaituri, Tommi (2000). "Automaattisten puhelinkeskusten historia" (http://www.cs.helsinki.fi/u/kerola/tkhist/k2000/alustukset/puhelinkeskukset/) (in Finnish). . Retrieved 21 March 2009.

[29] Palmberg, Christopher; Martikainen, Olli (23 May 2003). "Overcoming a Technological Discontinuity – The Case of the Finnish Telecom Industry and the GSM" (http://www.etla.fi/files/677_dp855.pdf) (PDF). The Research Institute of the Finnish Economy. . Retrieved 14 June 2009.

[30] "Puolustusvoimat: Kalustoesittely – Sanomalaitejärjestelmä" (http://www.mil.fi/maavoimat/kalustoesittely/index.dsp?level=81) (in Finnish). The Finnish Defence Forces. 15 June 2005. . Retrieved 14 May 2008.

[31] "The Finnish Defence Forces: Presentation of equipment: Message device" (http://www.mil.fi/maavoimat/kalustoesittely/00030_en.dsp). The Finnish Defence Forces. . Retrieved 14 June 2009.

[32] "Nokia 1100 phone offers reliable and affordable mobile communications for new growth markets" (http://press.nokia.com/PR/200308/915317_5.html) (Press release). Nokia Corporation. 27 August 2003. . Retrieved 26 May 2009.

[33] "Nokia – Mobira Cityman – The move to mobile – Story of Nokia" (http://www.nokia.com/about-nokia/company/story-of-nokia/the-move-to-mobile/mobira-cityman). Nokia Corporation. . Retrieved 14 May 2008.

[34] Juutilainen, Matti. "Siirtyvä tietoliikenne, luennot 7–8: Matkapuhelinverkot" (http://www.it.lut.fi/kurssit/06-07/Ti5312600/luentokalvot/luento07-08.pdf) (in Finnish) (PDF). Lappeenranta University of Technology. . Retrieved 22 March 2009.

[35] "Nokia – Mobile era begins – The move to mobile – Story of Nokia" (http://www.nokia.com/about-nokia/company/story-of-nokia/the-move-to-mobile/mobile-era-begins). Nokia Corporation. . Retrieved 20 March 2009.

[36] Karttunen, Anu (2 May 2003). "Tähdet syöksyvät, Benefon" (http://www.talouselama.fi/sijoittaminen/article165594.ece) (in Finnish). *Talouselämä* (Talentum Oyj). . Retrieved 28 July 2009.

[37] "Nokia´s Pioneering GSM Research and Development to be Awarded by Eduard Rhein Foundation" (http://press.nokia.com/1997/10/17/nokiaÂ´s-pioneering-gsm-research-and-development-to-be-awarded-by-eduard-rhein-foundation) (Press release). Nokia Corporation. 17 October 1997. . Retrieved 7 April 2012.

[38] "Global Mobile Communication is 20 years old" (http://gsmworld.com/newsroom/press-releases/2070.htm) (Press release). GSM Association. 6 September 2007. . Retrieved 23 March 2009.

[39] "Happy 20th birthday, GSM" (http://news.zdnet.co.uk/leader/0,1000002982,39289154,00.htm). *ZDNet.co.uk* (CBS Interactive). 7 September 2007. . Retrieved 23 March 2009.

[40] "Nokia – First GSM call – The move to mobile – Story of Nokia" (http://www.nokia.com/about-nokia/company/story-of-nokia/the-move-to-mobile/first-gsm-call). Nokia Corporation. . Retrieved 20 March 2009.

[41] Smith, Tony (9 November 2007). "15 years ago: the first mass-produced GSM phone" (http://www.reghardware.co.uk/2007/11/09/ft_nokia_1011/). *Register Hardware*. Situation Publishing Ltd. . Retrieved 23 March 2009.

[42] "Nokia – Nokia Tune – Mobile revolution – Story of Nokia" (http://www.nokia.com/about-nokia/company/story-of-nokia/mobile-revolution/nokia-tune). Nokia Corporation. . Retrieved 23 March 2009.

[43] "3 Billion GSM Connections On The Mobile Planet – Reports The GSMA" (http://www.gsmworld.com/newsroom/press-releases/2008/1108.htm). GSM Association. 16 April 2008. . Retrieved 21 March 2009.

[44] "Nokia MikroMikko 1" (http://www.old-computers.com/museum/computer.asp?st=1&c=630). Old-Computers.com. . Retrieved 14 May 2008.

[45] "Net – Fujitsun asiakaslehti, Net-lehden historia: 1980-luku" (http://www.fujitsuservices.fi/historia/net/1980.htm) (in Finnish). Fujitsu Services Oy, Finland. . Retrieved 22 March 2009.

[46] "Historia: 1991–1999" (http://www.fujitsu.com/fi/about/history/1991/) (in Finnish). Fujitsu Services Oy, Finland. . Retrieved 22 March 2009.

[47] Hietanen, Juha (28 February 2000). "Closure of Fujitsu Siemens plant – a repeat of Renault Vilvoorde?" (http://www.eiro.eurofound.eu.int/2000/02/feature/fi0002136f.html). EIRO, European Industrial Relations Observatory on-line. . Retrieved 14 May 2008.

[48] Hietanen, Juha (28 February 2000). "Fujitsu Siemens tehdas suljetaan – toistuiko Renault Vilvoord?" (http://www.eurofound.europa.eu/eiro/2000/02/word/fi0002136ffi.doc) (in Finnish) (DOC). EIRO, European Industrial Relations Observatory on-line. . Retrieved 14 May 2008.

[49] "ViewSonic Corporation Acquires Nokia Display Products' Branded Business" (http://press.nokia.com/PR/200001/775025_5.html) (Press release). Nokia Corporation. 17 January 2000. . Retrieved 22 March 2009.

[50] "Nokia Booklet 3G brings all day mobility to the PC world" (http://www.nokia.com/press/press-releases/showpressrelease?newsid=1336683) (Press release). Nokia Corporation. 24 August 2009. . Retrieved 26 August 2009.

[51] Pietilä, Antti-Pekka (27 September 2000). "Kari Kairamon nousu ja tuho" (http://www.taloussanomat.fi/arkisto/2000/09/27/kari-kairamon-nousu-ja-tuho/200026243/12) (in Finnish). *Taloussanomat*. . Retrieved 21 March 2009.

[52] "Finland: How bad policies turned bad luck into a recession" (http://www.cepr.org/PRESS/EP29 finland.htm). Centre for Economic Policy Research. . Retrieved 5 April 2009.

[53] Häikiö, Martti; translated by Hackston, David (2001) (in Finnish). *Nokia Oyj:n historia 1–3 (A history of Nokia plc 1–3)* (http://www.finlit.fi/booksfromfinland/bff/102/nokia.htm). Helsinki: Edita. ISBN 951-37-3467-6. . Retrieved 21 March 2008.

[54] "Nokia – Leading the world – Mobile revolution – Story of Nokia" (http://www.nokia.com/about-nokia/company/story-of-nokia/mobile-revolution/leading-the-world). Nokia Corporation. . Retrieved 21 March 2009.

[55] Reinhardt, Andy (3 August 2006). "Nokia's Magnificent Mobile-Phone Manufacturing Machine" (http://www.businessweek.com/globalbiz/content/aug2006/gb20060803_618811.htm). *BusinessWeek Online Europe*. . Retrieved 21 March 2009.

[56] Professor Voomann, Thomas E.; Cordon, Carlos (1998). "Nokia Mobile Phones: Supply Line Management" (http://stuff.mit.edu/afs/athena/course/15/15.795/Nokia Supply Chain Case Study.pdf) (PDF). Lausanne, Switzerland: IMD – International Institute for Management Development. . Retrieved 21 March 2009.

[57] Ewing, Jack (30 July 2007). "Why Nokia Is Leaving Moto in the Dust" (http://www.businessweek.com/magazine/content/07_31/b4044050.htm). *BusinessWeek Online*. . Retrieved 21 March 2009.

[58] Lin, Porter; Khan, Raedeep; Piekute, Vaida; Luhtasela, Jussi; Fang, Debby (1 December 2005). "Supply Chain Management Case Nokia" (http://imba.nccu.edu.tw/OIP/EXchange/Docs/F04/mis/final/group6/SCM in Nokia - Written report-V1.0.pdf) (PDF). IMBA, College of Commerce, National Chengchi University. . Retrieved 21 March 2009.

[59] Virki, Tarmo (5 March 2007). "Nokia's cheap phone tops electronics chart" (http://www.reuters.com/article/technologyNews/idUSL0262945620070503). Reuters. . Retrieved 14 May 2008.

[60] Saylor, Michael (2012). *The Mobile Wave: How Mobile Intelligence Will Change Everything*. Perseus Books/Vanguard Press. p. 81.

[61] "Nokia World 2007: Nokia outlines its vision of Internet evolution and commitment to environmental sustainability" (http://www.nokia.com/A4136001?newsid=1172937) (Press release). Nokia Corporation. 4 December 2007. . Retrieved 14 May 2008.

[62] "Nokia Productions and Spike Lee premiere the world's first social film" (http://www.nokia.com/press/press-releases/showpressrelease?newsid=1259528) (Press release). Nokia Corporation. 14 October 2008. . Retrieved 12 June 2009.

[63] "Nokia seizes social internet and amplifies music experience" (http://www.nokia.com/press/press-releases/showpressrelease?newsid=1338896) (Press release). Nokia Corporation. 2 September 2009. . Retrieved 12 October 2009.

[64] "Nokia 7705 Twist launched Stateside on Verizon (photo gallery)" (http://conversations.nokia.com/2009/09/10/nokia-7705-twist-launched-stateside-on-verizon-photo-gallery/). Nokia Corporation. 10 September 2009. . Retrieved 11 October 2009.

[65] "Nokia launches two new handsets under 'Asha' range" (http://economictimes.indiatimes.com/tech/hardware/nokia-launches-two-new-handsets-under-asha-range/articleshow/15421139.cms). 09-08-2012. .

[66] "Home of the Maemo community" (http://maemo.org/). maemo.org. . Retrieved 12 November 2011.

[67] Paul, Ryan (25 June 2010). "Nokia picks MeeGo Linux, not Symbian, for flagship phones" (http://arstechnica.com/open-source/news/2010/06/nokia-to-use-meego-linux-and-not-symbian-for-flagship-phones.ars). Ars Technica. . Retrieved 12 November 2011.

[68] Sherwood, James (13 August 2009). "Nokia exec denies Symbian Maemo swap claim" (http://www.reghardware.co.uk/2009/08/13/nokia_denies_maemo/). reghardware. . Retrieved 12 November 2011.

[69] "Nokia announces strategic partnership with Microsoft, will use WP7 as primary OS" (http://www.techit.in/2011/02/nokia-announces-strategic-partnership-with-microsoft-will-use-wp7-as-primary-os/). TechIt.in. .

[70] "Missed the historic Nokia+Microsoft event today? See it here!" (http://www.techit.in/2011/02/missed-the-historic-nokiamicrosoft-event-today-see-it-here/). TechIt.in. .

[71] "Nokia and Microsoft form partnership" (http://www.bbc.co.uk/news/business-12427680). BBC. 11 February 2011. . Retrieved 12 February 2011.

[72] "RIP: Symbian" (http://www.engadget.com/2011/02/11/rip-symbian/). Engadget. .

[73] "Capitulation: Nokia adopts Windows Phone 7" (http://arstechnica.com/gadgets/news/2011/02/nokia-adopts-windows-phone-7-as-primary-platform.ars). ArsTechnica. 11 February 2011. . Retrieved 12 February 2011.

[74] "Nokia will be able to customize 'everything' in Windows Phone 7, but likely won't" (http://www.engadget.com/2011/02/11/nokia-will-be-able-to-customize-everything-in-windows-phone-7). Engadget. .

[75] ben-Aaron, Diana (11 February 2011). "Nokia Falls Most Since July 2009 After Microsoft Deal" (http://www.bloomberg.com/news/2011-02-11/nokia-joins-forces-with-microsoft-to-challenge-dominance-of-apple-google.html). Bloomberg. .

[76] "Gartner Says Sales of Mobile Devices in Second Quarter of 2011 Grew 16.5 Percent Year-on-Year; Smartphone Sales Grew 74 Percent" (http://www.gartner.com/it/page.jsp?id=1764714) (Press release). Gartner. 11 August 2011. . Retrieved 29 September 2011.

[77] Ward, Andrew (21 July 2011). "Apple overtakes Nokia in smartphone stakes" (http://www.ft.com/cms/s/0/4d7fd1e2-b38e-11e0-b56c-00144feabdc0.html#axzz1SlVal4Cs). *Financial Times*. . Retrieved 21 July 2011.

[78] Fried, Ina (9 August 2011). "Nokia to Exit Symbian, Low-End Phone Businesses in North America" (http://allthingsd.com/20110809/exclusive-nokia-to-exit-symbian-low-end-phone-businesses-in-north-america/). All Things Digital. . Retrieved 9 August 2011.

[79] "Nokia's Lumia phones to show Groupon offers on maps" (http://www.bbc.com/news/technology-19093769). BBC. .

[80] Mobile operators unconvinced by Nokia's revival bid | Reuters (http://uk.reuters.com/article/2012/04/17/uk-nokia-telcos-idUKBRE83G08Z20120417)

[81] Chris Smith (30 June 2012). "Nokia promises back-up plan if Windows Phone fails" (http://www.techradar.com/news/phone-and-communications/mobile-phones/nokia-promises-back-up-plan-if-windows-phone-fails-1087600). Techradar. .

[82] "Nokia's Siilasmaa: Goal to regain competitiveness" (http://yle.fi/uutiset/nokias_siilasmaa_goal_to_regain_competitiveness/6199219). YLS Uutiset. 28 June 2012. .

[83] "Hungarian and Finnish Prime Ministers Inaugurate Nokia's "Factory of the Future" in Komárom" (http://press.nokia.com/PR/200005/780293_5.html) (Press release). Nokia Corporation. 5 May 2000. . Retrieved 22 March 2009.

[84] "Nokia to set up a new mobile device factory in Romania" (http://www.nokia.com/A4136001?newsid=1114420) (Press release). Nokia Corporation. 26 March 2007. . Retrieved 14 May 2008.

[85] "Nokia to open cell phone plant near Cluj" (http://web.archive.org/web/20080507194303/http://www.boston.com/news/world/europe/articles/2007/03/22/nokia_to_open_cell_phone_plant_near_cluj/). Associated Press. Boston.com. 22 March 2007. Archived from the original (http://www.boston.com/news/world/europe/articles/2007/03/22/nokia_to_open_cell_phone_plant_near_cluj/) on 7 May 2008. . Retrieved 14 May 2008.

[86] "Nokia to build mobile phone plant in Romania" (http://www.hs.fi/english/article/Nokia+to+build+mobile+phone+plant+in+Romania/1135226144930). *Helsingin Sanomat*. 27 March 2007. . Retrieved 14 May 2008.

[87] "German Politicians Return Cell Phones Amid Nokia Boycott Calls" (http://www.dw-world.de/dw/article/0,2144,3076534,00.html). *Deutsche Welle*. 18 January 2008. . Retrieved 22 March 2009.

[88] "German State Demands €60 Million from Nokia" (http://www.spiegel.de/international/business/0,1518,540699,00.html). *Der Spiegel*. 11 March 2008. . Retrieved 22 March 2009.

[89] "Nokia Networks takes strong measures to reduce costs, improve profitability and strengthen leadership position" (http://press.nokia.com/PR/200304/898905_5.html) (Press release). Nokia Corporation. 10 April 2003. . Retrieved 14 May 2008.

[90] "Nokia Networks to shed 1,800 jobs worldwide; majority of impact felt in Finland" (http://www2.hs.fi/english/archive/news.asp?id=20030411IE6). *Helsingin Sanomat*. 11 April 2003. . Retrieved 14 May 2008.

[91] Leyden, John (10 April 2003). "Nokia Networks axes 1,800 staff" (http://www.theregister.co.uk/2003/04/10/nokia_networks_axes/). *The Register*. . Retrieved 14 May 2008.

[92] "Nokia's Law (transcription)" (http://www.yle.fi/mot/kj050117/englishscript.htm). YLE TV1, Mot. 17 January 2005. . Retrieved 14 May 2008.

[93] "Nokia and Sanyo proposed new company will not proceed" (http://www.nokia.com/A4136002?newsid=1059331) (Press release). Nokia Corporation. 26 June 2006. . Retrieved 14 May 2008.

[94] "Nokia decides not to go forward with Sanyo CDMA partnership and plans broad restructuring of its CDMA business" (http://www.nokia.com/A4136002?newsid=1059329) (Press release). Nokia Corporation. 22 June 2006. . Retrieved 14 May 2008.

[95] "Nokia and Sanyo Announce Intent to Form a Global CDMA Mobile Phones Business" (http://www.nokia.com/A4136002?newsid=1034612) (Press release). Nokia Corporation. 14 February 2006. . Retrieved 14 May 2008.

[96] "Shell appoints Jorma Ollila as new Chairman" (http://www.shell.com/home/content/media/news_and_library/press_releases/2005/pr_announcement_04082005.html) (Press release). Royal Dutch Shell. 4 August 2005. . Retrieved 22 March 2009.

[97] "Nokia moves forward with management succession plan" (http://www.nokia.com/A4136002?newsid=1004430) (Press release). Nokia Corporation. 1 August 2005. . Retrieved 22 March 2009.

[98] Repo, Eljas; Melender, Tommi (19 September 2005). "Changing the guard at Nokia – Olli-Pekka Kallasvuo takes the helm" (http://finland.fi/netcomm/news/showarticle.asp?intNWSAID=41296&LAN=ENG). *Ministry for Foreign Affairs of Finland*. Virtual Finland. . Retrieved 22 March 2009.

[99] Kallasvuo, Olli-Pekka; President and CEO (8 May 2008). "2008 Nokia Annual General Meeting (transcription)" (http://nds1.nokia.com/NOKIA_COM_1/Microsites/AGM_2008/pdf/OPK_AGM_2008_ENGLISH.pdf) (PDF). Helsinki Fair Centre, Amfi Hall: Nokia Corporation. . Retrieved 12 June 2009.

[100] " □□□□□□□□ " (http://www.nokia.co.jp/about/release_081127.shtml) (in Japanese). – –. 27 November 2008.. Retrieved 5 December 2008.

[101] Nokia Will Lay off 4,000 and Move More Manufacturing to Asia | PCWorld Business Center (http://www.pcworld.com/businesscenter/article/249507/nokia_will_lay_off_4000_and_move_more_manufacturing_to_asia.html)

[102] Nokia Lays Off 1,000 Employees From Finnish Plant, Will Focus On Software (http://i2mag.com/nokia-lays-off-1000-employees-from-finnish-plant-will-focus-on-software/)

[103] "Nokia completes acquisition of assets of Sega.com Inc." (http://www.nokia.com/A4136002?newsid=918198) (Press release). Nokia Corporation. 22 September 2003.. Retrieved 16 March 2009.

[104] "Nokia to extend leadership in enterprise mobility with acquisition of Intellisync" (http://www.nokia.com/A4136002?newsid=1021663) (Press release). Nokia Corporation. 16 November 2005.. Retrieved 22 March 2009.

[105] "Nokia completes acquisition of Intellisync" (http://www.nokia.com/A4136002?newsid=1034184) (Press release). Nokia Corporation. 10 February 2006.. Retrieved 22 March 2009.

[106] "Nokia and Siemens to merge their communications service provider businesses" (http://www.nokia.com/A4136002?newsid=1057716) (Press release). Nokia Corporation. 19 June 2006.. Retrieved 22 March 2009.

[107] "Nokia to acquire Loudeye and launch a comprehensive mobile music experience" (http://www.nokia.com/A4136002?newsid=1067845) (Press release). Nokia Corporation. 8 August 2006.. Retrieved 14 May 2008.

[108] "Nokia completes Loudeye acquisition" (http://www.nokia.com/A4136001?newsid=1081455) (Press release). Nokia Corporation. 16 October 2006.. Retrieved 14 May 2008.

[109] "Nokia acquires Twango to offer a comprehensive media sharing experience" (http://www.nokia.com/A4136001?newsid=1141417) (Press release). Nokia Corporation. 24 July 2007.. Retrieved 14 May 2008.

[110] "Nokia Acquires Twango – Frequently Asked Questions (FAQ)" (http://www.nokia.com/NOKIA_COM_1/Press/Materials/NokiaTwangoFAQ.pdf) (PDF). Nokia Corporation.. Retrieved 14 May 2008.

[111] "Nokia to acquire Enpocket to create a global mobile advertising leader" (http://www.nokia.com/A4136001?newsid=1153772) (Press release). Nokia Corporation. 17 September 2007.. Retrieved 14 May 2008.

[112] Niccolai, James (1 October 2007). "Nokia buys mapping service for $8.1 billion" (http://www.infoworld.com/article/07/10/01/Nokia-buys-mapping-service-for-8.1-billion_1.html). *IDG News Service* (InfoWorld).. Retrieved 14 May 2008.

[113] "Nokia completes its acquisition of NAVTEQ" (http://www.nokia.com/A4136001?newsid=1235107) (Press release). Nokia Corporation. 10 July 2008.. Retrieved 22 March 2009.

[114] "Nokia to acquire leading consumer email and instant messaging provider OZ Communications" (http://news.taume.com/World-Business/Business-Finance/Nokia-to-acquire-leading-consumer-email-and-instant-messaging-provider-OZ-Communications-6922). *Taume News*. 30 September 2008.. Retrieved 30 September 2008.

[115] "Nokia to acquire cellity" (http://www.nokia.com/press/press-releases/showpressrelease?newsid=1330831) (Press release). Nokia Corporation. 24 July 2009.. Retrieved 4 August 2009.

[116] "Nokia completes acquisition of cellity" (http://www.nokia.com/press/press-releases/showpressrelease?newsid=1332884) (Press release). Nokia Corporation. 5 August 2009.. Retrieved 6 August 2009.

[117] "Nokia has acquired Plum" (http://www.nokia.com/press/press-releases/showpressrelease?newsid=1340931) (Press release). Nokia Corporation. 11 September 2009.. Retrieved 28 January 2010.

[118] "Nokia Acquires Browser Firm Novarra" (http://techie-buzz.com/mobile-news/nokia-acquires-browser-firm-novarra.html). Techie-buzz. 28 March 2010.. Retrieved 29 March 2010.

[119] "Nokia Acquires MetaCarta" (http://www.informationweek.com/news/mobility/smart_phones/showArticle.jhtml?articleID=224202519). informationweek. 11 April 2010.. Retrieved 12 April 2010.

[120] http://www.theverge.com/mobile/2012/1/7/2690366/nokia-buys-smarterphone-developer-of-feature-phone-operating-system

[121] http://www.pocket-lint.com/news/46116/nokia-acquires-scalado-for-better-image-quality

[122] "Nokia to cut 3500 jobs worldwide; to shut Romania factory" (http://www.moneycontrol.com/news/world-news/nokia-to-cut-3500-jobs-worldwide-to-shut-romania-factory_592235.html). 29 September 2011..

[123] Moen, Arild (8 February 2012). "Nokia to Cut 4,000 Jobs" (http://online.wsj.com/article/SB10001424052970204136404577210401816583074.html). *The Wall Street Journal*..

[124] "Nokia to cut 10,000 jobs; shut units" (http://www.thehindu.com/business/companies/article3528038.ece). *The Hindu*. 14 June 2012..

[125] ben-Aaron, Diana (14 June 2012). "Nokia to Cut 10,000 Jobs as Elop Tries to Stanch Losses" (http://www.bloomberg.com/news/2012-06-14/nokia-to-cut-10-000-jobs-as-elop-tries-to-stanch-losses.html). *Bloomberg*..

[126] http://timesofindia.indiatimes.com/tech/itslideshow/14151500.cms

[127] http://i.nokia.com/blob/view/-/263824/data/1/-/Request-Nokia-in-2010-pdf.pdf

[128] Nokia Downgraded to Junk - Zacks.com (http://www.zacks.com/stock/news/77208/nokia-downgraded-to-junk)

[129] "Nokia CEO Stephen Elop admits failure to foresee fast-changing industry" (http://economictimes.indiatimes.com/news/international-business/nokia-ceo-stephen-elop-admits-failure-to-foresee-fast-changing-industry/articleshow/14466105.cms). 28 June 2012..

[130] "Structure" (http://www.nokia.com/about-nokia/company/structure). Nokia Corporation. 1 October 2009.. Retrieved 28 December 2009.

[131] "Nokia Siemens Networks starts operations and assumes a leading position in the communications industry" (http://www.nokia.com/A4136002?newsid=1116423) (Press release). Nokia Corporation. 2 April 2007.. Retrieved 7 April 2009.

[132] "Nokia's 25 percent profit jump falls short of expectations" (http://www.usatoday.com/money/economy/2008-04-17-173945271_x.htm). Associated Press. USA Today. 17 April 2008. . Retrieved 14 May 2008.
[133] "Worldwide Mobile Phone Market Maintains Its Growth Trajectory in the Fourth Quarter Despite Soft Demand for Feature Phones, According to IDC" (http://www.idc.com/getdoc.jsp?containerId=prUS23297412). ICD. 1 February 2012. .
[134] "The Wave of the Future" (http://www.underconsideration.com/brandnew/archives/the_wave_of_the_future.php). *Brand New: Opinions on Corporate and Brand Identity Work*. UnderConsideration LLC. 25 March 2007. . Retrieved 14 May 2008.
[135] "Reviews – 2007 – Nokia Siemens Networks" (http://www.identityworks.com/reviews/2007/Nokia_Siemens.htm). *Identityworks*. 2007. . Retrieved 14 May 2008.
[136] "Nokia Leadership Team" (http://www.nokia.com/A4126335). Nokia Corporation. April 2007. . Retrieved 14 May 2008.
[137] "Board of Directors" (http://www.nokia.com/A4126350). Nokia Corporation. April 2007. . Retrieved 14 May 2008.
[138] "Audit Committee Charter at Nokia" (http://www.nokia.com/NOKIA_COM_1/About_Nokia/Sidebars_new_concept/Board_charters/audit_charter.pdf) (PDF). Nokia Corporation. 2007. . Retrieved 14 May 2008.
[139] "Personnel Committee Charter at Nokia" (http://www.nokia.com/NOKIA_COM_1/About_Nokia/Sidebars_new_concept/Board_charters/personnel_charter_2007.pdf) (PDF). Nokia Corporation. 2007. . Retrieved 14 May 2008.
[140] "Corporate Governance and Nomination Committee Charter at Nokia" (http://www.nokia.com/NOKIA_COM_1/About_Nokia/Sidebars_new_concept/Board_charters/CG_Charter_2008_Final_20080123.pdf) (PDF). Nokia Corporation. 2008. . Retrieved 14 May 2008.
[141] "Committees of the Board" (http://www.nokia.com/link?cid=EDITORIAL_4207). Nokia Corporation. May 2007. . Retrieved 14 May 2008.
[142] Virkkunen, Johannes (29 September 2006). "New Finnish Companies Act designed to increase Finland's competitiveness" (http://www.lmr.fi/publications/companies_act_290906.pdf) (PDF). *LMR Attorneys Ltd. (Luostarinen Mettälä Räikkönen)*. . Retrieved 14 May 2008.
[143] "Articles of Association" (http://www.nokia.com/NOKIA_COM_1/About_Nokia/Company/Corporate_Governance/Articles_of_Association/Nokia_Articles_of_Association_10052007.pdf) (PDF). Nokia Corporation. 10 May 2007. . Retrieved 14 May 2008.
[144] "Corporate Governance Guidelines at Nokia" (http://www.nokia.com/NOKIA_COM_1/About_Nokia/Sidebars_new_concept/Board_charters/corporate_governance_guideline_sep06.pdf) (PDF). Nokia Corporation. 2006. . Retrieved 14 May 2008.
[145] "Suomalaisten yritysten ylin johto" (http://www.kolumbus.fi/taglarsson/dokumentit/yritys.htm) (in Finnish). . Retrieved 20 March 2009.
[146] "Nokia Research Center" (http://www.nokia.com/NOKIA_COM_1/Press/twwln/press_kit/Nokia_Research_Center_Press_Backgrounder_October_2007.pdf) (PDF). Nokia Corporation. October 2007. . Retrieved 14 May 2008.
[147] "About NRC – Nokia Research Center" (http://research.nokia.com/aboutus/index.html). Nokia Corporation. . Retrieved 17 March 2009.
[148] "NRC Locations – Nokia Research Center" (http://research.nokia.com/locations/index.html). Nokia Corporation. . Retrieved 17 March 2009.
[149] "INdT – Instituto Nokia de Tecnologia" (http://www.indt.org.br/). Nokia Corporation. . Retrieved 17 March 2009.
[150] "Production units" (http://www.nokia.com/A4149133). Nokia Corporation. June 2008. . Retrieved 14 May 2008.
[151] "Nokia despre sechestrul ANAF: Colaborăm pentru a ne asigura că situaţia e soluţionată satisfăcător" (http://www.mediafax.ro/economic/nokia-despre-sechestrul-anaf-colaboram-pentru-a-ne-asigura-ca-situatia-e-solutionata-satisfacator-8963720) (Press release). mediafax.ro. 12 November 2011. . Retrieved 12 November 2011.
[152] Kapanen, Ari (24 July 2007). "Ulkomaalaiset valtaavat pörssiyhtiöitä" (http://www.taloussanomat.fi/porssi-ja-raha/2007/07/24/Ulkomaalaiset+valtaavat+pörssiyhtiöitä/200717658/103) (in Finnish). *Taloussanomat*. . Retrieved 14 May 2008.
[153] "Nokia is no longer Finland's most valuable company" (http://www.phonearena.com/news/Nokia-is-no-longer-Finlands-most-valuable-company_id28750). phonearena.com. 4 April 2012. .
[154] Ali-Yrkkö, Jyrki (2001). "The role of Nokia in the Finnish Economy" (http://www.etla.fi/files/940_FES_01_1_nokia.pdf) (PDF). ETLA (The Research Institute of the Finnish Economy). . Retrieved 21 March 2009.
[155] Ali-Yrkkö, Jyrki (2010). "NOKIA AND FINLAND IN A SEA OF CHANGE" (http://www.etla.fi/files/2585_nokia_kirja_8_v2_kansineen.pdf). *ETLA – Research Institute of the Finnish Economy*. . Retrieved 12 November 2011.
[156] Guardian newspaper: Nokia cuts 4,000 jobs and moves smartphone manufacturing to Asia, 9 February 2012 (http://www.guardian.co.uk/technology/2012/feb/08/nokia-cuts-4000-manufacturing-jobs)
[157] "HS Archives" (http://www.hs.fi/arkisto/haku?pageNumber=1&order=FIFO&advancedSearch=&free=Connecting+People+Ove+Strandberg&date=year2003&depa=Kaikki+osastot&fromDay=0&fromMonth=0&fromYear=0&toDay=0&toMonth=0&toYear=0) (in Finnish). Helsingin Sanomat. 1 June 2003. . Retrieved 14 May 2008.
[158] "NOKIA | Connecting Pople 1992 Vector Logo (AI EPS)" (http://www.hdicon.com/vector-logos/nokia-connecting-pople-1992/). *HDicon.com*. . Retrieved 17 October 2010.
[159] Pitkänen, Juhani (Nokia's Art Director) (3 September 2007). "Nokia Strategic Marketing, Brand Identity" (http://www.nokia.com/A4126575). Nokia Corporation. . Retrieved 14 May 2008.
[160] "NOKIA | Connecting Pople new Vector Logo (AI EPS)" (http://www.hdicon.com/vector-logos/nokia-connecting-pople-new/). *HDicon.com*. . Retrieved 17 October 2010.
[161] "Erik Spiekermann – Furniture, Designs & Home Decor" (http://www.dwr.com/category/designers/r-t/erik-spiekermann.do). Design Within Reach. . Retrieved 7 January 2010.
[162] "Our New Typeface" (http://brandbook.nokia.com/blog/view/item62250/). Nokia Little Blog of Branding. . Retrieved 3 April 2012.

[163] http://thenextweb.com/eu/2012/04/04/poor-nokia-isnt-even-the-most-valuable-company-in-finland-anymore/
[164] http://media.corporate-ir.net/media_files/IROL/10/107224/Nokia_results2011Q2e.pdf
[165] http://www.theregister.co.uk/2012/04/19/nokia_earnings_ouch/
[166] http://thenextweb.com/insider/2012/05/14/nokia-is-getting-pummeled-stock-price-hits-staggering-low-after-a-6-nosedive/
[167] http://yle.fi/uutiset/nokia_share_price_in_nose_dive_to_2_euros/6094843
[168] "Nokia Way and values" (http://www.nokia.com/A4126303). Nokia Corporation. . Retrieved 14 May 2008.
[169] "dotMobi Investors" (http://web.archive.org/web/20080507214157/http://mtld.mobi/company/about/investors). dotMobi. Archived from the original (http://mtld.mobi/company/about/investors) on 7 May 2008. . Retrieved 14 May 2008.
[170] Haumont, Serge; Siren, Ritva. "dotMobi, a Key Enabler for the Mobile Internet" (http://research.nokia.com/files/Haumont-dotMobi.pdf) (PDF). *Nokia Research Center*. Nokia Corporation. . Retrieved 14 May 2008.
[171] "Nokia Ad Business" (http://www.adservice.nokia.com/faq.jsp#12). Nokia Corporation. . Retrieved 14 May 2008.
[172] Reardon, Marguerite (6 March 2007). "Nokia introduces mobile ad services" (http://news.com.com/2100-1039_3-6164800.html). *CNET News.com*. . Retrieved 14 May 2008.
[173] "Meet Ovi, the door to Nokia's Internet services" (http://www.nokia.com/A4136001?newsid=1149749) (Press release). Nokia Corporation. 29 August 2007. . Retrieved 7 April 2009.
[174] Niccolai, James (4 December 2007). "Nokia Lays Plan for More Internet Services" (http://www.nytimes.com/idg/IDG_002570DE00740E18002573A70046F2EF.html?ref=technology). *IDG News Service* (New York Times). . Retrieved 14 May 2008.
[175] "Ovi by Nokia" (http://www.nokia.com/NOKIA_COM_1/Press/Materials/White_Papers/pdf_files/backgrounders2008/Backgrounder_Ovi_by_Nokia.pdf) (PDF). Nokia Corporation. . Retrieved 7 April 2009.
[176] "Ovi Store opens for business" (http://www.nokia.com/press/press-releases/showpressrelease?newsid=1317441) (Press release). Nokia Corporation. 26 May 2009. . Retrieved 12 June 2009.
[177] Virki, Tarmo (18 March 2009). "Nokia to shutter its "Mosh" success story" (http://www.reuters.com/article/technologyNews/idUSTRE52H6AI20090318?pageNumber=1&virtualBrandChannel=0). *Reuters*. . Retrieved 14 July 2009.
[178] "Nokia launches new online magazine" (http://www.newstatesman.com/magazines/2010/03/online-magazine-nokia-ovi). NewStatesman. 23 March 2010. . Retrieved 29 March 2010.
[179] "Nokia – My Nokia" (http://europe.nokia.com/my-nokia). Nokia Corporation. . Retrieved 10 January 2012.
[180] "Nokia Retreats from Music Service" (http://www.yle.fi/uutiset/news/2011/01/nokia_retreats_from_music_service_2294706.html). YLE. 18 January 2011. . Retrieved 18 January 2011.
[181] Fields, Davis (17 December 2008). "Nokia Email service graduates as part of Nokia Messaging" (http://betalabs.nokia.com/blog/2008/12/17/nokia-email-service-graduates-as-part-of-nokia-messaging/). *Nokia Beta Labs*. Nokia Corporation. . Retrieved 16 March 2009.
[182] "Nokia Messaging: FAQ" (http://email.nokia.com/account/faq.action?change_locale=en). Nokia Corporation. . Retrieved 12 June 2009.
[183] Cellan-Jones, Rory (22 June 2009). "Hi-tech helps Iranian monitoring" (http://news.bbc.co.uk/1/hi/technology/8112550.stm). *BBC News*. . Retrieved 14 July 2009.
[184] Rhoads, Christopher; Chao, Loretta (22 June 2009). "Iran's Web Spying Aided By Western Technology" (http://online.wsj.com/article/SB124562668777335653.html#mod). *The Wall Street Journal* (Dow Jones & Company, Inc.): pp. A1. . Retrieved 14 July 2009.
[185] "Provision of Lawful Intercept capability in Iran" (http://www.nokiasiemensnetworks.com/global/Press/Press+releases/news-archive/Provision+of+Lawful+Intercept+capability+in+Iran.htm) (Press release). Nokia Siemens Networks. 22 June 2009. . Retrieved 14 July 2009.
[186] Kamali Dehghan, Saeed (14 July 2009). "Iranian consumers boycott Nokia for 'collaboration'" (http://www.guardian.co.uk/world/2009/jul/14/nokia-boycott-iran-election-protests). *The Guardian* (London: Guardian News and Media Limited). . Retrieved 27 July 2009.
[187] Ozimek, John (6 March 2009). "'Lex Nokia' company snoop law passes in Finland" (http://www.theregister.co.uk/2009/03/06/finland_nokia_snooping/). *The Register*. . Retrieved 27 July 2009.
[188] "Nokia Denies Threat to Leave Finland" (http://www.cellular-news.com/story/35783.php). *cellular-news*. 1 February 2009. . Retrieved 27 July 2009.
[189] Virki, Tarmo (18 January 2010). "SCENARIOS-What lies ahead in Nokia vs Apple legal battle" (http://www.reuters.com/article/idUSLDE60H05R20100118?type=marketsNews). *Reuters*. . Retrieved 25 January 2010.
[190] "The war of the Smartphones: Nokia's new patent suit against Apple" (http://pda-phone-reviews.in/latest-news/nokias-new-patent-suit-against-apple/). *Snartphone Reviews*. 6 January 2010. . Retrieved 25 January 2010.
[191] "Nokia's Patent Settlement With Apple Won't Help Much" (http://www.informationweek.com/news/personal-tech/smart-phones/230600172). 14 Jun 2011. . Retrieved 29 June 2011.
[192] Smith, Catharine (14 Jun 2011). "Apple Settles With Nokia In Patent Lawsuit" (http://www.huffingtonpost.com/2011/06/14/apple-nokia-patent-lawsuit-settlement_n_876499.html). *Huffington Post*. . Retrieved 29 June 2011.
[193] ben-Aaron, Diana; Pohjanpalo, Kati (14 Jun 2011). "Nokia Wins Apple Patent-License Deal Cash, Settles Lawsuits" (http://www.bloomberg.com/news/2011-06-14/nokia-apple-payments-to-nokia-settle-all-litigation.html). *Bloomberg*. . Retrieved 29 June 2011.
[194] "Guide to Greener Electronics – Greenpeace International" (http://www.greenpeace.org/international/en/campaigns/climate-change/cool-it/Guide-to-Greener-Electronics/). Greenpeace International. . Retrieved 14 November 2011.
[195] "Nokia – Where and how to recycle – Recycling – Environment" (http://www.nokia.com/environment/recycling/where-and-how-to-recycle). Nokia. . Retrieved 12 August 2010.

[196] "Global consumer survey reveals that majority of old mobile phones are lying in drawers at home and not being recycled" (http://www.nokia.com/press/press-releases/showpressrelease?newsid=1234291) (Press release). Nokia Corporation. 8 July 2008. . Retrieved 27 July 2009.

[197] "Nokia – Energy efficiency – Devices and services – Environment" (http://www.nokia.com/environment/we-energise/nokia-and-energy-efficiency). Nokia. . Retrieved 12 August 2010.

[198] "Nokia – Energy saving targets – Environmental strategy – Strategy and reports – Environment" (http://www.nokia.com/environment/strategy-and-reports/environmental-strategy/energy-saving-targets). Nokia. . Retrieved 12 August 2010.

[199] "Nokia" (http://www.greenpeace.org/international/en/campaigns/toxics/electronics/Guide-to-Greener-Electronics/companies/Nokia/). Greenpeace International. . Retrieved 12 August 2010.

[200] "Materials and substances" (http://www.nokia.com/environment/we-create/materials-and-substances). Nokia Corporation. . Retrieved 12 August 2010.

[201] "Eco declarations" (http://www.nokia.com/environment/we-create/devices-and-accessories/eco-declarations). Nokia Corporation. . Retrieved 27 July 2009.

[202] Rubio, Jenalyn (12 April 2008). "Tech Goes Greener" (http://www.pcworld.com/article/id,144482-c,recycling/article.html). *Computerworld Philippines* (PC World). . Retrieved 14 May 2008.

[203] "Nokia Remade Concept Phone goes Green" (http://www.mobiletor.com/2008/04/09/nokia-remade-concept-phone-goes-green/). Mobiletor. 9 April 2008. . Retrieved 14 May 2008.

[204] "Open Innovation – Nokia Research Center" (http://research.nokia.com/openinnovation). Nokia Corporation. . Retrieved 1 April 2009.

[205] "Nokia is India's most trusted brand : Brand Trust" (http://profit.ndtv.com/news/show/nokia-is-india-s-most-trusted-brand-brand-trust-136783). profit.ndtv.com. 19 January 2011. . Retrieved 21 November 2011.

Further reading

Title	Author	Publisher	Year	Length	ISBN
Winning Across Global Markets: How Nokia Creates Strategic Advantage in a Fast-Changing World	Dan Steinbock	Jossey-Bass / Wiley	May 2010	304 pp	ISBN 978-0-470-33966-4
Nokia: The Inside Story	Martti Häikiö	FT / Prentice Hall	October 2002	256 pp	ISBN 0-273-65983-9
Work Goes Mobile: Nokia's Lessons from the Leading Edge	Michael Lattanzi, Antti Korhonen, Vishy Gopalakrishnan	John Wiley & Sons	January 2006	212 pp	ISBN 0-470-02752-5
Mobile Usability: How Nokia Changed the Face of the Mobile Phone	Christian Lindholm, Turkka Keinonen, Harri Kiljander	McGraw-Hill Companies	June 2003	301 pp	ISBN 0-07-138514-2
Business The Nokia Way: Secrets of the World's Fastest Moving Company	Trevor Merriden	John Wiley & Sons	February 2001	168 pp	ISBN 1-84112-104-5
The Nokia Revolution: The Story of an Extraordinary Company That Transformed an Industry	Dan Steinbock	AMACOM Books	April 2001	375 pp	ISBN 0-8144-0636-X

External links

- Official Nokia international website (http://www.nokia.com/)

Touchscreen

A **touchscreen** is an electronic visual display that can detect the presence and location of a touch within the display area. The term generally refers to touching the display of the device with a finger or hand. Touchscreens can also sense other passive objects, such as a stylus. Touchscreens are common in devices such as game consoles, all-in-one computers, tablet computers, and smartphones.

A child solves a computerized puzzle using a touchscreen.

The touchscreen has two main attributes. First, it enables one to interact directly with what is displayed, rather than indirectly with a pointer controlled by a mouse or touchpad. Secondly, it lets one do so without requiring any intermediate device that would need to be held in the hand (other than a stylus, which is optional for most modern touchscreens). Such displays can be attached to computers, or to networks as terminals. They also play a prominent role in the design of digital appliances such as the personal digital assistant (PDA), satellite navigation devices, mobile phones, and video games.

iPad tablet computer on a stand

The popularity of smartphones, tablet computers and many types of information appliances is driving the demand and acceptance of common touchscreens for portable and functional electronics. With a display of a simple smooth surface, and direct interaction without any hardware (keyboard or mouse) between the user and content, fewer accessories are required. Touchscreens are popular in the medical field, and in heavy industry, as well as kiosks such as museum displays or room automation, where keyboard and mouse systems do not allow a suitably intuitive, rapid, or accurate interaction by the user with the display's content.

Historically, the touchscreen sensor and its accompanying controller-based firmware have been made available by a wide array of after-market system integrators, and not by display, chip, or motherboard manufacturers. Display manufacturers and chip manufacturers worldwide have acknowledged the trend toward acceptance of touchscreens as a highly desirable user interface component and have begun to integrate touchscreens into the fundamental design of their products.

History

E.A. Johnson described his work on capacitive touch screens in a short article published in 1965[5] and then more fully—along with photographs and diagrams—in an article published in 1967.[6] A description of the applicability of the touch technology for air traffic control was described in an article published in 1968.[7] Bent Stumpe with the aid of Frank Beck, both engineers from CERN, developed a transparent touch screen in the early 1970s and it was manufactured by CERN and put to use in 1973.[8] A resistive touch screen was developed by American inventor G Samuel Hurst and the first version produced in 1982.[9]

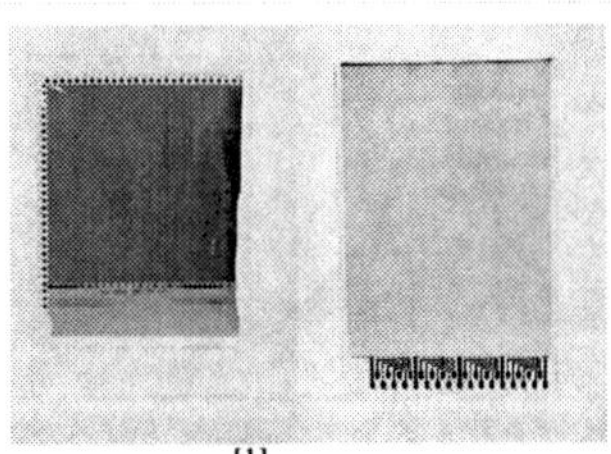

The prototype[1] x-y mutual capacitance touchscreen (left) developed at CERN[2][3] in 1977 by Bent Stumpe, a Danish electronics engineer, for the control room of CERN's accelerator SPS (Super Proton Synchrotron). This was a further development of the self-capacitance screen (right), also developed by Stumpe at CERN[4] in 1972.

From 1979–1985, the Fairlight CMI (and Fairlight CMI IIx) was a high-end musical sampling and re-synthesis workstation that utilized light pen technology, with which the user could allocate and manipulate sample and synthesis data, as well as access different menus within its OS by touching the screen with the light pen. The later Fairlight series IIT models used a graphics tablet in place of the light pen. The HP-150 from 1983 was one of the world's earliest commercial touchscreen computers. Similar to the PLATO IV system, the touch technology used employed infrared transmitters and receivers mounted around the bezel of its 9" Sony Cathode Ray Tube (CRT), which detected the position of any non-transparent object on the screen.

In 1986 the first graphical point of sale software was demonstrated on the 16-bit Atari 520ST color computer. It featured a color touchscreen widget-driven interface.[10] The ViewTouch[11] point of sale software was first shown by its developer, Gene Mosher, at Fall Comdex, 1986, in Las Vegas, Nevada to visitors at the Atari Computer demonstration area and was the first commercially available POS system with a widget-driven color graphic touch screen interface.

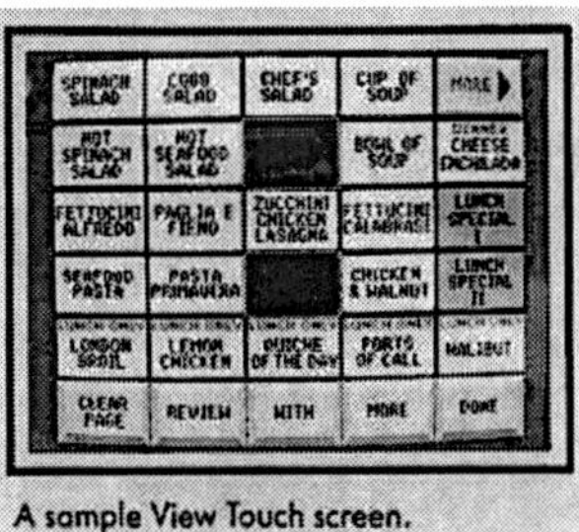

ViewTouch POS widget-driven touch screen GUI

An early attempt at a handheld game console with touchscreen controls was Sega's intended successor to the Game Gear, though the device was ultimately shelved and never released due to the expensive cost of touchscreen technology in the early 1990s. Touchscreens would not be popularly used for video games until the release of the Nintendo DS in 2004.[12] Until recently, most consumer touchscreens could only sense one point of contact at a time, and few have had the capability to sense how hard one is touching. This has changed with the commercialization of multi-touch technology.

Technologies

There are a variety of touchscreen technologies that have different methods of sensing touch.

Resistive

A resistive touchscreen panel comprises several layers, the most important of which are two thin, transparent electrically-resistive layers separated by a thin space. These layers face each other, with a thin gap between.The top screen (the screen that is touched) has a coating on the underside surface of the screen. Just beneath it is a similar resistive layer on top of its substrate. One layer has conductive connections along its sides, the other along top and bottom. A voltage is passed through one layer, and sensed at the other. When an object, such as a fingertip or stylus tip, presses down on the outer surface, the two layers touch to become connected at that point: The panel then behaves as a pair of voltage dividers, one axis at a time. By rapidly switching between each layer, the position of a pressure on the screen can be read.

Resistive touch is used in restaurants, factories and hospitals due to its high resistance to liquids and contaminants. A major benefit of resistive touch technology is its low cost. Disadvantages include the need to press down, and a risk of damage by sharp objects. Resistive touchscreens also suffer from poorer contrast, due to having additional reflections from the extra layer of material placed over the screen.

Surface acoustic wave

Surface acoustic wave (SAW) technology uses ultrasonic waves that pass over the touchscreen panel. When the panel is touched, a portion of the wave is absorbed. This change in the ultrasonic waves registers the position of the touch event and sends this information to the controller for processing. Surface wave touchscreen panels can be damaged by outside elements. Contaminants on the surface can also interfere with the functionality of the touchscreen.[13]

Capacitive

Capacitive touchscreen of a mobile phone

A capacitive touchscreen panel consists of an insulator such as glass, coated with a transparent conductor such as indium tin oxide (ITO).[14][15] As the human body is also an electrical conductor, touching the surface of the screen results in a distortion of the screen's electrostatic field, measurable as a change in capacitance. Different technologies may be used to determine the location of the touch. The location is then sent to the controller for processing. Unlike a resistive touchscreen, one cannot use a capacitive touchscreen through most types of electrically insulating material, such as gloves; one requires a special capacitive stylus, or a special-application glove with an embroidered patch of conductive thread passing through it and contacting the user's fingertip. This disadvantage especially affects usability in consumer electronics, such as touch tablet PCs and capacitive smartphones in cold weather. In response, there are gloves on the market with one or more fingertips that include conductive material, thus overcoming this limitation. The largest capacitive display manufacturers continue to develop thinner and more accurate touchscreens, with LCD touchscreens for mobile devices now being produced with 'in-cell' technology that eliminates a layer, by building the capacitors inside the LCD itself. This type of touchscreen reduces the visible distance (within millimetres) between the user's finger and what the user is touching on the screen, creating a more direct contact with the content displayed and enabling taps and gestures to be even more responsive.

Surface capacitance

In this basic technology, only one side of the insulator is coated with a conductive layer. A small voltage is applied to the layer, resulting in a uniform electrostatic field. When a conductor, such as a human finger, touches the uncoated surface, a capacitor is dynamically formed. The sensor's controller can determine the location of the touch indirectly from the change in the capacitance as measured from the four corners of the panel. As it has no moving parts, it is moderately durable but has limited resolution, is prone to false signals from parasitic capacitive coupling, and needs calibration during manufacture. It is therefore most often used in simple applications such as industrial controls and kiosks.[16]

Projected capacitance

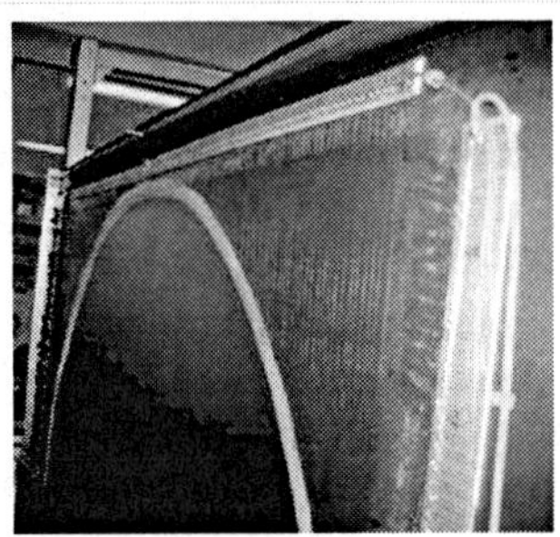

Back side of a Multitouch Globe, based on Projected Capacitive Touch (PCT) technology.

Projected Capacitive Touch (PCT; also PCAP) technology is a variant of capacitive touch technology. All PCT touch screens are made up of a matrix of rows and columns of conductive material, layered on sheets of glass. This can be done either by etching a single conductive layer to form a grid pattern of electrodes, or by etching two separate, perpendicular layers of conductive material with parallel lines or tracks to form a grid. Voltage applied to this grid creates a uniform electrostatic field, which can be measured. When a conductive object, such as a finger, comes into contact with a PCT panel, it distorts the local electrostatic field at that point. This is measurable as a change in capacitance. If a finger bridges the gap between two of the "tracks," the charge field is further interrupted and detected by the controller. The capacitance can be changed and measured at every individual point on the grid (intersection). Therefore, this system is able to accurately track touches.[17] Due to the top layer of a PCT being glass, it is a more robust solution than less costly resistive touch technology. Additionally, unlike traditional capacitive touch technology, it is possible for a PCT system to sense a passive stylus or gloved fingers. However, moisture on the surface of the panel, high humidity, or collected dust can interfere with the performance of a PCT system. There are two types of PCT: mutual capacitance and self-capacitance.

Mutual capacitance

This is common PCT approach, which makes use of the fact that most conductive objects are able to hold a charge if they are very close together. In mutual capacitive sensors, there is a capacitor at every intersection of each row and each column. A 16-by-14 array, for example, would have 224 independent capacitors. A voltage is applied to the rows or columns. Bringing a finger or conductive stylus close to the surface of the sensor changes the local electrostatic field which reduces the mutual capacitance. The capacitance change at every individual point on the grid can be measured to accurately determine the touch location by measuring the voltage in the other axis. Mutual capacitance allows multi-touch operation where multiple fingers, palms or styli can be accurately tracked at the same time.

Self-capacitance

Self-capacitance sensors can have the same X-Y grid as mutual capacitance sensors, but the columns and rows operate independently. With self-capacitance, the capacitive load of a finger is measured on each column or row electrode by a current meter. This method produces a stronger signal than mutual capacitance, but it is unable to resolve accurately more than one finger, which results in "ghosting", or misplaced location sensing.

Infrared

An infrared touchscreen uses an array of X-Y infrared LED and photodetector pairs around the edges of the screen to detect a disruption in the pattern of LED beams. These LED beams cross each other in vertical and horizontal patterns. This helps the sensors pick up the exact location of the touch. A major benefit of such a system is that it can detect essentially any input including a finger, gloved finger, stylus or pen. It is generally used in outdoor applications and point of sale systems which can not rely on a conductor (such as a bare finger) to activate the touchscreen. Unlike capacitive touchscreens, infrared touchscreens do not require any patterning on the glass which increases durability and optical clarity of the overall system. Infrared touchscreens are sensitive to dirt/dust that can interfere with the IR beams, and suffer from parallax in curved surfaces and accidental press when the user hovers his/her finger over the screen while searching for the item to be selected.

Infrared sensors mounted around the display watch for a user's touchscreen input on this PLATO V terminal in 1981. The monochromatic plasma display's characteristic orange glow is illustrated.

Optical imaging

Optical touchscreens are a relatively modern development in touchscreen technology, in which two or more image sensors are placed around the edges (mostly the corners) of the screen. Infrared back lights are placed in the camera's field of view on the other side of the screen. A touch shows up as a shadow and each pair of cameras can then be pinpointed to locate the touch or even measure the size of the touching object (see visual hull). This technology is growing in popularity, due to its scalability, versatility, and affordability, especially for larger units.

Dispersive signal technology

Introduced in 2002 by 3M, this system uses sensors to detect the piezoelectricity in the glass that occurs due to a touch. Complex algorithms then interpret this information and provide the actual location of the touch.[18] The technology claims to be unaffected by dust and other outside elements, including scratches. Since there is no need for additional elements on screen, it also claims to provide excellent optical clarity. Also, since mechanical vibrations are used to detect a touch event, any object can be used to generate these events, including fingers and stylus. A downside is that after the initial touch the system cannot detect a motionless finger.

Acoustic pulse recognition

In this system, introduced by Tyco International's Elo division in 2006, the key to the invention is that a touch at each position on the glass generates a unique sound. Four tiny transducers attached to the edges of the touchscreen glass pick up the sound of the touch. The sound is then digitized by the controller and compared to a list of prerecorded sounds for every position on the glass. The cursor position is instantly updated to the touch location. APR is designed to ignore extraneous and ambient sounds, since they do not match a stored sound profile. APR differs from other attempts to recognize the position of touch with transducers or microphones, in using a simple table lookup method rather than requiring powerful and expensive signal processing hardware to attempt to calculate the touch location without any references.[19] The touchscreen itself is made of ordinary glass, giving it good durability and optical clarity. It is usually able to function with scratches and dust on the screen with good accuracy. The technology is also well suited to displays that are physically larger. Similar to the dispersive signal technology system, after the initial touch, a motionless finger cannot be detected. However, for the same reason, the touch recognition is not disrupted by any resting objects.

Construction

There are several principal ways to build a touchscreen. The key goals are to recognize one or more fingers touching a display, to interpret the command that this represents, and to communicate the command to the appropriate application.

In the most popular techniques, the capacitive or resistive approach, there are typically four layers:

1. Top polyester coated with a transparent metallic conductive coating on the bottom
2. Adhesive spacer
3. Glass layer coated with a transparent metallic conductive coating on the top
4. Adhesive layer on the backside of the glass for mounting.

When a user touches the surface, the system records the change in the electrical current that flows through the display.

Dispersive-signal technology which 3M created in 2002, measures the piezoelectric effect—the voltage generated when mechanical force is applied to a material—that occurs chemically when a strengthened glass substrate is touched.

There are two infrared-based approaches. In one, an array of sensors detects a finger touching or almost touching the display, thereby interrupting light beams projected over the screen. In the other, bottom-mounted infrared cameras record screen touches.

In each case, the system determines the intended command based on the controls showing on the screen at the time and the location of the touch.

Development

Most touchscreen patents were filed during the 1970s and 1980s and have expired. Touchscreen component manufacturing and product design are no longer encumbered by royalties or legalities with regard to patents and the use of touchscreen-enabled displays is widespread.

The development of multipoint touchscreens facilitated the tracking of more than one finger on the screen; thus, operations that require more than one finger are possible. These devices also allow multiple users to interact with the touchscreen simultaneously.

With the growing use of touchscreens, the marginal cost of touchscreen technology is routinely absorbed into the products that incorporate it and is nearly eliminated. Touchscreens now have proven reliability. Thus, touchscreen displays are found today in airplanes, automobiles, gaming consoles, machine control systems, appliances, and handheld display devices including the Nintendo DS and multi-touch enabled cellphones; the touchscreen market for mobile devices is projected to produce US$5 billion in 2009.[20]

The ability to accurately point on the screen itself is also advancing with the emerging graphics tablet/screen hybrids.

TapSense, announced in October 2011, allows touchscreens to distinguish what part of the hand was used for input, such as the fingertip, knuckle and fingernail. This could be used in a variety of ways, for example, to copy and paste, to capitalize letters, to active different drawing modes, and similar.[21][22]

Ergonomics and usage

Fingernail as stylus

These ergonomic issues of direct touch can be bypassed by using a different technique, provided that the user's fingernails are either short or sufficiently long. Rather than pressing with the soft skin of an outstretched fingertip, the finger is curled over, so that the tip of a fingernail can be used instead. This method does not work on capacitive touchscreens.

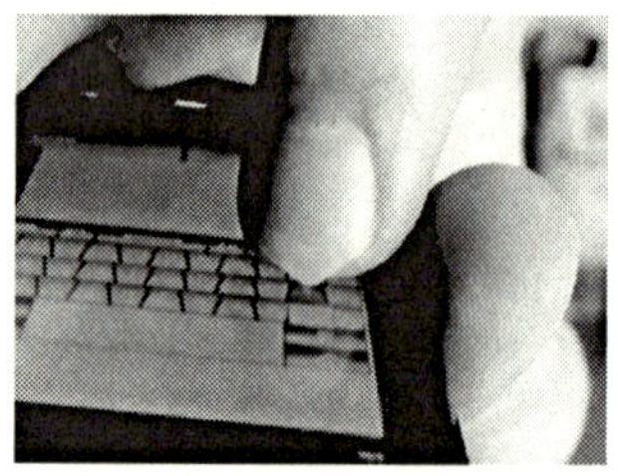

Pointed nail for easier typing. This concept--using a pointed fingernail to specifically use as a stylus for writing-tablet communication--appeared in the 1950 science fiction short story "Scanners Live in Vain".

The fingernail's hard, curved surface contacts the touchscreen at one very small point. Therefore, much less finger pressure is needed, much greater precision is possible (approaching that of a stylus, with a little experience), much less skin oil is smeared onto the screen, and the fingernail can be silently moved across the screen with very little resistance, allowing for selecting text, moving windows, or drawing lines.

The human fingernail consists of keratin which has a hardness and smoothness similar to the tip of a stylus (and so will not typically scratch a touchscreen). Alternatively, very short stylus tips are available, which slip right onto the end of a finger; this increases visibility of the contact point with the screen.

Fingerprints

Touchscreens can suffer from the problem of fingerprints on the display. This can be mitigated by the use of materials with optical coatings designed to reduce the visible effects of fingerprint oils, or oleophobic coatings as used in the iPhone 3G S, which lessen the actual amount of oil residue, or by installing a matte-finish anti-glare screen protector, which creates a slightly roughened surface that does not easily retain smudges, or by reducing skin contact by using a fingernail or stylus.

Combined with haptics

Touchscreens are often used with haptic response systems. An example of this technology would be a system that caused the device to vibrate when a button on the touchscreen was tapped. The user experience with touchscreens lacking tactile feedback or haptics can be difficult due to latency or other factors. Research from the University of Glasgow Scotland [Brewster, Chohan, and Brown 2007 and more recently Hogan] demonstrates that sample users reduce input errors (20%), increase input speed (20%), and lower their cognitive load (40%) when touchscreens are combined with haptics or tactile feedback [vs. non-haptic touchscreens].

"Gorilla arm"

The Jargon File dictionary of hacker slang defined *"gorilla arm"* as the failure to understand the ergonomics of vertically mounted touchscreens for prolonged use. By this proposition the human arm held in an unsupported horizontal position rapidly becomes fatigued and painful, the so-called "gorilla arm".[23] It is often cited as a prima facie example of what not to do in ergonomics. Vertical touchscreens still dominate in applications such as ATMs and data kiosks in which the usage is too brief to be an ergonomic problem.

Discomfort might be caused by previous poor posture and atrophied muscular systems caused by limited physical exercise.[24]

Screen protectors

Some touchscreens, primarily those employed in smartphones, use transparent plastic protectors to prevent any scratches that might be caused by day-to-day use from becoming permanent.

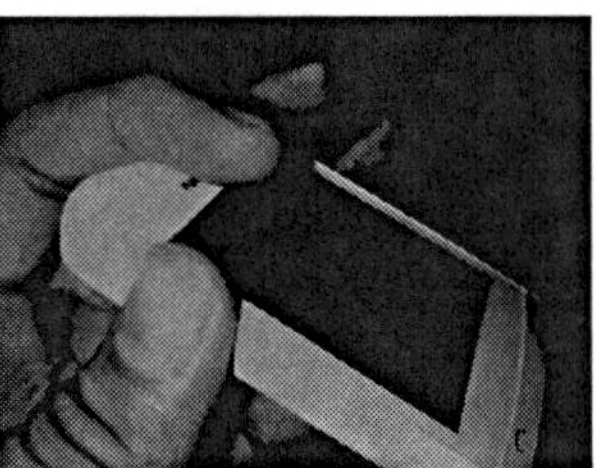

Installing a screen protector on a Nokia C2-02

Notes

[1] *The first capacitative touch screens at CERN* (http://cerncourier.com/cws/article/cern/42092), CERN Courrier, 31 March 2010, , retrieved 2010-05-25

[2] Bent STUMPE (16 March 1977), *A new principle for x-y touch system* (http://cdsweb.cern.ch/record/1266588/files/StumpeMar77.pdf), CERN, , retrieved 2010-05-25

[3] Bent STUMPE (6 February 1978), *Experiments to find a manufacturing process for an x-y touch screen* (http://cdsweb.cern.ch/record/1266589/files/StumpeFeb78.pdf), CERN, , retrieved 2010-05-25

[4] Frank BECK & Bent STUMPE (24 May 1973), *Two devices for operator interaction in the central control of the new CERN accelerator* (http://cdsweb.cern.ch/record/186242/files/p1.pdf), CERN, , retrieved 2010-05-25

[5] Johnson, E.A. (1965). "Touch Display - A novel input/output device for computers". *Electronics Letters* **1** (8): 219–220.

[6] Johnson, E.A. (1967). "Touch Displays: A Programmed Man-Machine Interface". *Ergonomics* **10** (2): 271–277.

[7] Orr, N.W.; Hopkins, V.D. (1968). "The Role of Touch Display in Air Traffic Control.". *The Controller* **7**: 7–9.

[8] CERN Bulletin Issue 12/2010 & 13/2010, "Another of CERN's many inventions!" (http://cdsweb.cern.ch/record/1248908)

[9] oakridger.com, "G. Samuel Hurst -- the 'Tom Edison' of ORNL", December 14 2010 (http://www.oakridger.com/columnists/x860698067/G-Samuel-Hurst-the-Tom-Edison-of-ORNL?zc_p=1).

[10] The ViewTouch restaurant system (http://www.atarimagazines.com/startv2n6/gettingdowntobusiness.html) by Giselle Bisson

[11] http://www.viewtouch.com

[12] Travis Fahs (April 21, 2009). "IGN Presents the History of SEGA" (http://retro.ign.com/articles/974/974695p7.html). IGN. p. 7. . Retrieved 2011-04-27.

[13] Patschon, Mark (1988-03-15). *Acoustic touch technology adds a new input dimension* (http://rwservices.no-ip.info:81/pens/biblio88.html#Platshon88). Computer Design. pp. 89–93.

[14] Kable, Robert G. (1986-07-15). United States Patent 4,600,807. http://rwservices.no-ip.info:81/pens/biblio86.html#Kable86

[15] Kable, Robert G. (1986-07-15). *Electrographic Apparatus* (http://www.freepatentsonline.com/4600807.pdf). United States Patent 4,600,807 (full image).

[16] "Please Touch! Explore The Evolving World Of Touchscreen Technology" (http://electronicdesign.com/Articles/Index.cfm?AD=1&ArticleID=18592). electronicdesign.com. . Retrieved 2009-09-02.

[17] Knowledge base: Multi-touch hardware (http://www.multi-touch-solution.com/en/knowledge-base-en/multitouch-technologies-en/)

[18] Beyers, Tim (2008-02-13). "Innovation Series: Touchscreen Technology" (http://www.fool.com/investing/general/2008/02/13/innovation-series-touchscreen-technology.aspx). *The Motley Fool.* . Retrieved 2009-03-16.

[19] *Acoustic Pulse Recognition Touchscreens* (http://media.elotouch.com/pdfs/marcom/apr_wp.pdf). Elo Touch Systems. 2006. p. 3. . Retrieved 2011-09-27

[20] "Touch Screens in Mobile Devices to Deliver $5 Billion Next Year | Press Release" (http://www.abiresearch.com/press/1231-Touch+Screens+in+Mobile+Devices+to+Deliver+$5+Billion+Next+Year). ABI Research. 2008-09-10. . Retrieved 2009-06-22.

[21] "New Screen Technology, TapSense, Can Distinguish Between Different Parts Of Your Hand" (http://techcrunch.com/2011/10/19/new-screen-technology-tapsense-can-distinguish-between-different-parts-of-your-hand/?utm_source=feedburner&utm_medium=feed&utm_campaign=Feed:+Techcrunch+(TechCrunch)). . Retrieved October 19, 2011.

[22] "TapSense: Enhancing Finger Interaction on Touch Surfaces" (http://www.chrisharrison.net/index.php/Research/TapSense/). . Retrieved 28 January 2012.

[23] "gorilla arm" (http://catb.org/jargon/html/G/gorilla-arm.html). Catb.org. . Retrieved 2012-01-04.

[24] USA (2011-10-03). "Poor posture subjects a worker's body to muscle imbalance, nerve compression, Langford ML, Occup Health Saf. 1994 Sep;63(9):38-40, 42" (http://www.ncbi.nlm.nih.gov/pubmed/9156441). Ncbi.nlm.nih.gov. . Retrieved 2012-01-04.

References

- "Touch screens now offer compelling uses". *IEEE Software* **8** (2): 93–94, 107. 1991. doi:10.1109/52.73754.
- Potter, R.; Weldon, L. & Shneiderman, B. (1988). Proc. CHI'88. Washington, DC: ACM Press. pp. 27–32.
- Sears, A.; Plaisant, C. & Shneiderman, B. (1992). "A new era for high precision touchscreens". In Hartson, R. & Hix, D.. *Advances in Human-Computer Interaction*. **3**. Ablex, NJ. pp. 1–33.
- Sears, A. & Shneiderman, B. (1991). "High precision touchscreen: Design strategies and comparison with a mouse". *Int. J. of Man-Machine Studies* **34** (4): 593–613. doi:10.1016/0020-7373(91)90037-8.
- Holzinger, A. (2003). "Finger Instead of Mouse: Touch Screens as a means of enhancing Universal Access". *In: Carbonell, N.; Stephanidis C. (Eds): Universal Access, Lecture Notes in Computer Science* **2615**: 387–397.

External links

- Howstuffworks (http://electronics.howstuffworks.com/question716.htm) - How do touchscreen monitors know where you're touching?
- MERL (http://diamondspace.merl.com/) - Mitsubishi Electric Research Lab (MERL)'s research on interaction with touch tables.
- Jefferson Y. Han et al. Multi-Touch Interaction Research (http://mrl.nyu.edu/~jhan/ftirtouch/). Multi-Input Touchscreen using Frustrated Total Internal Reflection.
- Dot-to-Dot Programming : Building Microcontrollers (http://www.popsci.com/diy/article/2009-01/dot-dot-programming)
- Knowledge base: Multi-touch hardware (http://www.multi-touch-solution.com/en/knowledge-base-en/multitouch-technologies-en/)
- Windows xp-7, Apple Mac & Lynux touchscreen software database : Software / Driver Download (http://www.touchscreensuk.com/)

Series_40

Series 40 is a software platform and application user interface (UI) software on Nokia's broad range of mid-tier feature phones, as well as on some of the Vertu line of luxury phones. It is the world's most widely used mobile phone platform and found in hundreds of millions of devices.[1] Nokia announced on 25 January 2012 that the company has sold over 1.5 billion S40 devices. [2] S40 has more features than the Series 30, which is a very basic OS. Neither of them are based on Symbian.

Series 40-based Nokia 6300

History

Series 40 was officially introduced in 1999 with the release of the Nokia 7110. It had a 96 × 65 pixel monochrome display and was the first phone to come with a WAP browser. Over the years, the S40 UI has evolved from a low resolution UI to a high resolution color UI with an enhanced graphical look. The third generation of Series 40 that became available in 2005 introduced support for devices with resolutions as high as QVGA (240×320).[3] It is possible to customize the look-and-feel of the UI via comprehensive themes.[4] A list of all Series 40 devices can be found on the Nokia web site.[5]

In 2012, the new Nokia Asha mobile phones 200/201, 302, 303 and 311 were released and all use Series 40.[5]

Home screen of Nokia S40 v10.80 running on Nokia 6303i

Features

Applications

It provides communication applications such as telephone, internet telephony (VoIP), messaging, email client with POP3 and IMAP4 capabilities and Web browser; media applications such as camera, video recorder, music/video player and FM radio; and phonebook and other personal information management (PIM) applications such as calendar and tasks. Basic file management, like in Series 60, is provided in the Applications and Gallery folders and subfolders. Gallery is also the default location for files transferred over Bluetooth to be placed. User-installed applications on Series 40 are generally mobile Java applications. Flash Lite applications are also supported, but mostly used for screensavers.[6]

Web browser

The integrated web browser can access most web content through the service provider's XHTML/HTML gateway. The latest version of Series 40, called Series 40 6th Edition, introduced a new browser based on the WebKit open source components WebCore and JavaScriptCore. The new browser delivers support for HTML 4.01, CSS2, JavaScript 1.5, and Ajax. Also, like the higher-end Series 60, Series 40 can run the Opera Mini web browser to enhance the user's web browsing experience.

Synchronization

Support for SyncML synchronization with external services of the address book, calendar and notes is present. However with many S40 phones, these synchronization settings must be sent via an OTA text message.

Technical

Software platform

Series 40 is an embedded software platform that is open for software development via standard or de-facto content and application development technologies. It supports Java MIDlets, i.e. Java MIDP and CLDC technology, which provide location, communication, messaging, media, and graphics capabilities.[7] S40 also supports Flash Lite applications.[6]

Operating system

Series 40 is a simpler operating system than the higher end S60. Because S40 devices do not support true multi-tasking and do not have a native code API for third parties, its user interface may appear to be more responsive and faster than the other Nokia platforms Symbian S60 on similar hardware.[8]

References

[1] "Forum Nokia - Nokia Series 40 Platform" (http://www.forum.nokia.com/Devices/Series_40/). Nokia. . Retrieved 2010-10-27.
[2] "Nokia has sold over 1.5 billion Series 40 phones" (http://www.esphoneblog.com/2012/01/25/nokia-has-sold-over-1-5-billion-series-40-phones/). . Retrieved 2012-01-25.
[3] "Series 40 UI Style Guide – Forum Nokia" (http://www.forum.nokia.com/info/sw.nokia.com/id/73e935fe-8b59-43b2-ab3e-1c5f763672db/Series_40_UI_Style_Guide.html). Nokia. . Retrieved 2008-09-26.
[4] "Carbide.ui Theme Edition (can be used to create S40 themes) – Forum Nokia" (http://www.forum.nokia.com/Library/Tools_and_downloads/Other/Carbide.ui/). Nokia. . Retrieved 2011-05-16.
[5] "Device specifications, filtered for Series 40" (http://www.developer.nokia.com/Devices/Device_specifications/?filter1=s40). Nokia. . Retrieved 2012-06-01.
[6] "Working with Nokia Series 40 Flash Lite content – Adobe Developer Center" (http://web.archive.org/web/20080518220200/http://www.adobe.com/devnet/devices/articles/nokia_series40_pt1_print.html). Adobe Systems. Archived from the original (http://www.adobe.com/devnet/devices/articles/nokia_series40_pt1_print.html) on 2008-05-18. . Retrieved 2008-09-26.
[7] "Developing Scalable Series 40 Applications, A Guide for Java Developers" (http://www.pearsoned.co.uk/BOOKSHOP/detail.asp?item=100000000075919). Addison-Wesley. . Retrieved 2008-09-26.
[8] "Comparing Series 40 against S60 (as of 2007) – All about Symbian" (http://www.allaboutsymbian.com/features/item/Series_40_vs_S60.php). . Retrieved 2008-09-26.

Graphics_display_resolution

Overview by vertical resolution and aspect ratio

Lines	5:4 = 1.25	4:3 = 1.3	3:2 = 1.5	16:10 = 1.6	5:3 = 1.6	16:9 = 1.7
120		160 QQVGA				
160			240 HQVGA			
240		320 QVGA		384 WQVGA	400 WQVGA	432 FWQVGA
320			480 HVGA			
360						640 nHD
480		640 VGA			800 WVGA	854 FWVGA
540						960 qHD
576						1024 WSVGA
600		800 SVGA			1024 WSVGA (17:10)	
640			960 DVGA	1024		
720					1152	1280 HD/WXGA
768		1024 XGA	1152		1280 WXGA	1366 WXGA
800				1280 WXGA		
864		1152 XGA+	1280			
900				1440 WXGA+		1600 HD+
960		1280 SXGA−	1440			
1024	1280 SXGA					
1050		1400 SXGA+		1680 WSXGA+		
1080						1920 FHD
1152						2048 QWXGA
1200		1600 UXGA		1920 WUXGA		
1440						2560 (W)QHD
1536		2048 QXGA				
1600				2560 WQXGA		
2048	2560 QSXGA		3200 WQSXGA (25:16)			
2160						3840 QFHD
2400		3200 QUXGA		3840 WQUXGA		
3072		4096 HXGA				
3200				5120 WHXGA		
4096	5120 HSXGA		6400 WHSXGA (25:16)			
4320						7680 UHD
4800		6400 HUXGA		7680 WHUXGA		

The **graphics display resolution** describes the width and height dimensions of a display, such as a computer monitor, in pixels. Certain combinations of width and height are standardized and typically given a name and an initialism that is descriptive of its dimensions. A higher display resolution means that displayed content appears sharper.

Aspect ratio

The gradual change of the favored aspect ratio of mass market display industry products, from 4:3, then to 16:10, and then to 16:9, has made many of the display resolutions listed in this article difficult to obtain in mass market products. The 4:3 aspect ratio generally reflects older products, especially the era of the cathode ray tube (CRT). The 16:10 aspect ratio had its largest use in the 1995–2010 period, and the 16:9 aspect ratio tends to reflect the newest (post 2010) mass market computer monitor, laptop, and entertainment products displays. In many cases the resolutions listed in the sections below may have a small market, may only be seen in specialized industrial or computer market products, or may not be available for sale.

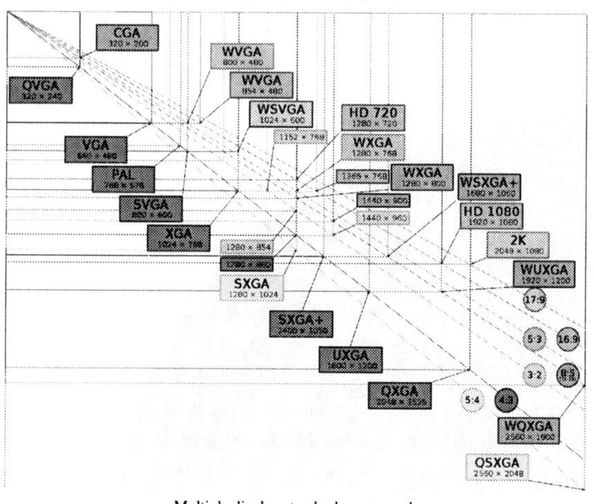

Multiple display standards compared.

The 4:3 aspect ratio was common in older television cathode ray tube (CRT) displays, which were not easily adaptable to a wider aspect ratio. When good quality alternate technologies (i.e., liquid crystal displays (LCDs) and plasma displays) became more available and less costly, around the year 2000, the common computer displays and entertainment products moved to a wider aspect ratio, first to the 16:10 ratio. The 16:10 ratio allowed some compromise between showing older 4:3 aspect ratio broadcast TV shows, but also allowing better viewing of widescreen movies. However, around the year 2005, entertainment industry displays (i.e., TV sets) gradually moved from 16:10 to the 16:9 aspect ratio, for further improvement of viewing widescreen movies. By about 2007, virtually all mass market entertainment displays were 16:9. In 2011, 1920×1080 was the favored resolution in the most heavily marketed entertainment market displays.

The computer display industry maintained the 16:10 aspect ratio longer than the entertainment industry, but in the 2005–2010 period, computers were increasingly marketed as dual use products, with uses in the traditional computer applications, but also as means of viewing entertainment content. In this time frame, almost all desktop, laptop, and display manufacturers gradually moved to promoting only 16:9 aspect ratio displays. By 2011, the 16:10 aspect ratio had virtually disappeared from the laptop display market. One artifact is that the highest available resolution in laptop displays moved downward in this time frame (i.e., the move from 1920×1200 laptop displays to 1920×1080 displays).

Video Graphics Array

Video Graphics Array

Name	x (px)	y (px)	x:y	x·y (Mpx)
QQVGA	160	120	4:3	0.019
HQVGA	240	160	3:2	0.038
QVGA	320	240	4:3	0.077
WQVGA	400	240	5:3	0.096
HVGA	480	320	3:2	0.154
VGA	640	480	4:3	0.307
WVGA	800	480	5:3	0.384
FWVGA	~854	480	16:9	0.410
SVGA	800	600	4:3	0.480
DVGA	960	640	3:2	0.614
WSVGA	1024	576	16:9	0.590
WSVGA	1024	600	17:10	0.614

QQVGA (160×120)

Quarter-QVGA (**QQVGA** or **qqVGA**) denotes a resolution of 160×120 or 120×160 pixels, usually used in displays of handheld devices. The term Quarter-QVGA signifies a resolution of one fourth the number of pixels in a QVGA display (half the number of vertical and half the number of horizontal pixels) which itself has one fourth the number of pixels in a VGA display.

The abbreviation *qqVGA* may be used to distinguish *quarter* from *quad*, just like *qVGA*.[1]

HQVGA (240×160)

Half-QVGA denotes a display screen resolution of 240×160 or 160×240 pixels, as seen on the Game Boy Advance. This resolution is half of QVGA, which is itself a quarter of VGA, which is 640×480 pixels.

QVGA (320×240)

The **Quarter Video Graphics Array** (also known as **Quarter VGA**, **QVGA**, or **qVGA**) is a popular term for a computer display with 320×240 display resolution. QVGA displays are most often used in mobile phones, personal digital assistants (PDA), and some handheld game consoles. Often the displays are in a "portrait" orientation (i.e., taller than they are wide, as opposed to "landscape") and are referred to as 240×320.[2]

QVGA compared to VGA

The name comes from having a **q**uarter of the 640×480 maximum resolution of the original IBM **VGA** display technology, which became a de facto industry standard in the late 1980s. QVGA is not a standard mode offered by the VGA BIOS, even though VGA and compatible chipsets support a QVGA-sized Mode X. The term refers only to the display's resolution and thus the abbreviated term QVGA or Quarter VGA is more appropriate to use.

QVGA resolution is also used in digital video recording equipment as a low-resolution mode requiring less data storage capacity than higher resolutions, typically in still digital cameras with video recording capability, and some mobile phones. Each frame is an image of 320×240 pixels. QVGA video is typically recorded at 15 or 30 frames per second. QVGA mode describes the size of an image in pixels, commonly called the resolution; numerous video file formats support this resolution.

While QVGA is a *lower* resolution than VGA, at higher resolutions the "Q" prefix commonly means *quad(ruple)* or four times *higher* display resolution (e.g., QXGA is 4 times higher resolution than XGA). To distinguish *quarter* from *quad*, lowercase "q" is sometimes used for "quarter" and uppercase "Q" for "quad", by analogy with SI prefixes like m/M and p/P, but this is not a consistent usage.[3]

Some examples of devices that use QVGA resolution include, Samsung i5500, Samsung Galaxy Fit, Samsung Galaxy Y, Samsung Galaxy Pocket, HTC Wildfire, Sony Ericsson Xperia X10 Mini, Sony Ericsson Xperia X10 Mini pro and Nintendo 3DS' bottom screen.

WQVGA (400×240)

Variants of WQVGA

x (px)	y (px)	x:y	x·y (Mpx)
376	240	4.7:3	0.0902
384	240	16:10	0.0922
400	240	15:9	0.0960
428	240	16:9	0.0103
432	240	16:9	0.0104
480	270	16:9	0.0130
480	272	16:9	0.0131

Wide QVGA or **WQVGA** is any display resolution having the same height in pixels as QVGA, but wider. This definition is consistent with other 'wide' versions of computer displays.

Since QVGA is 320 pixels wide and 240 pixels high (aspect ratio of 4:3), the resolution of a WQVGA screen might be 384×240 (8:5 aspect ratio), 400×240 (5:3—such as the Nintendo 3DS screen or the maximum resolution in YouTube at 240p), 428×240 or 432×240 (~16:9 ratio). As with WVGA, exact ratios of *n*:9 are not practical because of the way VGA controllers internally deal with pixels. For instance, when using graphical combinatorial operations on pixels, VGA controllers will use 1 bit per pixel. Since bits cannot be accessed individually but by chunks of 16 or an even higher power of 2, this limits the horizontal resolution to a 16-pixel granularity, i.e., the horizontal resolution must be divisible by 16. In the case of 16:9 ratio, with 240 pixels high, the horizontal resolution should be 240 / 9 × 16 = 426.6, the closest multiple of 16 is 432.

WQVGA has also been used to describe displays that are not 240 pixels high, for example Sixteenth HD1080 displays which are 480 pixels wide and 270 or 272 pixels high. This may be due to QVGA having the nearest screen height.

WQVGA resolutions are commonly used in touch screen mobile phones, such as 400×240, 432×240, and 480×240. For example, the Sony Ericsson Aino and the Samsung Instinct both have WQVGA screen resolutions—240×432. Other devices such as the Apple iPod nano also use a WQVGA screen, 240×376 pixels.

HVGA (480×320)

Variants of HVGA

x (px)	y (px)	x:y	x·y (Mpx)
480	270	16:9	0.1296
480	272	16:9	0.1306
480	320	3:2	0.1536
640	240	8:3	0.1536
480	360	4:3	0.1728

HVGA (**Half-size VGA**) screens have 480×320 pixels (3:2 aspect ratio), 480×360 pixels (4:3 aspect ratio), 480×272 (~16:9 aspect ratio) or 640×240 pixels (8:3 aspect ratio). The former is used by a variety of PDA devices, starting with the Sony CLIÉ PEG-NR70 in 2002, and standalone PDAs by Palm. The latter was used by a variety of handheld PC devices. VGA resolution is 640×480.

Examples of devices that use HVGA include the Apple iPhone 2G–3GS, BlackBerry Bold 9000, HTC Dream, Hero, Wildfire S, LG GW620 Eve, MyTouch 3G Slide, Nokia 6260 Slide, Palm Pre, Samsung M900 Moment, Sony Ericsson Xperia X8, Sony Ericsson Xperia mini, Sony Ericsson Xperia mini pro, Sony Ericsson Xperia active and Sony Ericsson Xperia Live with Walkman.

Texas Instruments produces the DLP pico projector which supports HVGA resolution.[4]

HVGA was the only resolution supported in the first versions of Google Android, up to release 1.5.[5] Other higher and lower resolutions are now available starting on release 1.6, like the popular WVGA resolution on the Motorola Droid or the QVGA resolution on the HTC Tattoo.

Three dimensional computer graphics common on television throughout the 1980s were mostly rendered at this resolution, causing objects to have jagged edges on the top and bottom when edges were not anti-aliased.

VGA (640×480)

Video Graphics Array (**VGA**) refers specifically to the display hardware first introduced with the IBM PS/2 line of computers in 1987,[6] but through its widespread adoption has also come to mean either an analog computer display standard, the 15-pin D-subminiature VGA connector or the 640×480 resolution itself. While this resolution was superseded in the personal computer market in the 1990s, it is becoming a popular resolution on mobile devices.[7] VGA is still the universal fallback troubleshooting mode in the case of trouble with graphic device drivers in operating systems.

WVGA (800×480)

Variants of WVGA

x (px)	y (px)	x:y	x·y (Mpx)
800	480	15:9	0.384
848	480	16:9	0.407
852	480	16:9	0.409
854	480	16:9	0.410

Wide VGA or **WVGA**, sometimes just **WGA**, an abbreviation for **Wide Video Graphics Array** is any display resolution with the same 480 pixel height as VGA but wider, such as 800×480 (aspect ratio 5:3), 848×480, 852×480 or 854×480 (~16:9). It is a common resolution among LCD projectors and later portable and hand-held internet-enabled devices (such as MID and Netbooks) as it is capable of rendering web sites designed for an 800 wide window in full page-width. Examples of hand-held internet devices, without phone capability, with this resolution include: ASUS Eee PC 700 series, Dell XCD35, Nokia 770, N800, and N810.

Mobile phones with WVGA display resolution are also common. A list of mobile phones with WVGA display is available.

FWVGA (854×480)

FWVGA is an abbreviation for **Full Wide Video Graphics Array** which refers to a display resolution of 854×480 pixels. 854×480 is approximately the 16:9 aspect ratio of anamorphically "un-squeezed" NTSC DVD widescreen video and considered a "safe" resolution that does not crop any of the image. It is called **Full WVGA** to distinguish it from other, narrower WVGA resolutions which require cropping 16:9 aspect ratio high-definition video (i.e. it is full width, albeit with considerable reduction in size). The 854 pixel width is rounded up from 853.3. $480 \times {}^{16}\!/_{9} = {}^{7680}\!/_{9} = 853{}^{1}\!/_{3}$. Since a pixel must be a whole number, rounding up to 854 ensures inclusion of the entire image.[8] Due to physical devices often being manufactured with pixel resolutions that are multiples of 16, the horizontal resolution of 854 may be implemented by the OS simply pretending the 10 edgemost columns, from a full physical width of 864, don't exist.

In 2010, mobile phones with FWVGA display resolution started to become more common. A list of mobile phones with FWVGA displays is available.

SVGA (800×600)

Super Video Graphics Array or **Ultra Video Graphics Array**,[9] almost always abbreviated to **Super VGA**, **Ultra VGA** or just **SVGA** or **UVGA** is a broad term that covers a wide range of computer display standards.[10]

Originally, it was an extension to the VGA standard first released by IBM in 1987. Unlike VGA—a purely IBM-defined standard—Super VGA was defined by the Video Electronics Standards Association (VESA), an open consortium set up to promote interoperability and define standards. When used as a resolution specification, in contrast to VGA or XGA for example, the term *SVGA* normally refers to a resolution of 800 × 600 pixels.

DVGA (960×640)

DVGA (**Double-size VGA**) screens have 960×640 pixels (3:2 aspect ratio). Both dimensions are double that of HVGA, hence the pixel count is quadrupled.

Examples of devices that use DVGA include Meizu MX mobile phone and the Apple iPhone 4/4S, where the screen is called "Retina Display".

WSVGA (1024×576/600)

The wide version of SVGA is known as **WSVGA** (**Wide Super VGA**), featured on Ultra-Mobile PCs, netbooks, and tablet computers. The resolution is either 1024×576 (aspect ratio 16:9) or 1024×600 (between 15:9 and 16:9) with screen sizes normally ranging from 7 to 10 inches.

Extended Graphics Array

Extended Graphics Array

Name	x (px)	y (px)	x:y	x·y (Mpx)
XGA	1024	768	4:3	0.786
WXGA	1280	720	16:9	0.922
WXGA	1280	768	5:3	0.983
WXGA	1280	800	16:10	1.024
WXGA	1360	768	~16:9	1.044
WXGA	1366	768	~16:9	1.049
XGA+	1152	864	4:3	0.995
WXGA+	1440	900	16:10	1.296
SXGA	1280	1024	5:4	1.310
SXGA+	1400	1050	4:3	1.470
WSXGA+	1680	1050	16:10	1.764
UXGA	1600	1200	4:3	1.920
WUXGA	1920	1200	16:10	2.304

XGA (1024×768)

XGA, the **Extended Graphics Array**, is an IBM display standard introduced in 1990. Later it became the most common appellation of the 1024×768 pixels display resolution, but the official definition is broader than that. It was not a new and improved replacement for Super VGA, but rather became one particular subset of the broad range of capabilities covered under the "Super VGA" umbrella.

The initial version of XGA expanded upon IBM's VGA, adding support for two resolutions:

- 800×600 pixels with high color (16 bits per pixel; i.e. 65,536 colors).
- 1024×768 pixels with a palette of 256 colors (8 bits per pixel)

Like its predecessor (the IBM 8514), XGA offered **fixed function** hardware acceleration to offload processing of 2D drawing tasks. XGA and 8514 could offload line-draw, bitmap-copy (bitblt), and color-fill operations from the host CPU. XGA's acceleration was faster than 8514's, and more comprehensive in that it supported more drawing primitives and XGA's 16 bits per pixel (65,536 color) display-mode.

XGA-2 added Truecolor mode for 640×480, high color mode and higher refresh rates for 1024×768, and improved accelerator performance. All XGA modes have a 4:3 aspect ratio rounded to 8 pixels.

XGA should not be confused with **EVGA** (Extended Video Graphics Array), a contemporaneous VESA standard that also has 1024×768 pixels.

WXGA (1280×768)

Variants of WXGA

x (px)	y (px)	x:y	x·y (Mpx)
1280	720	16:9	0.922
1280	768	15:9	0.983
1280	800	16:10	1.024
1360	768	16:9	1.044
1366	768	16:9	1.049

Wide Extended Graphics Array (**Wide XGA** or **WXGA**) is a set of non standard resolutions derived from the XGA display standard by widening it to a wide screen aspect ratio. WXGA is commonly used for low-end LCD TVs and LCD computer monitors for widescreen presentation.

When referring to televisions and other monitors intended for consumer entertainment use, WXGA is generally understood to refer to a resolution of 1366 (1365.333)×768,[11] with an aspect ratio of 16:9. In 2006 this was the most popular resolution for liquid crystal display televisions while XGA was for Plasma TVs flat panel displays.[12]

When referring to laptop displays or monitors intended primarily as computer displays, WXGA is most commonly used to refer to a resolution of 1280×800 pixels with an aspect ratio of 16:10.[13][14][15] This resolution is particularly popular for most laptops with a 14" or 15" screen. The exact resolution this refers to is somewhat variable, however, as the 1280x*nnn* resolutions were among the first widescreen resolutions commonly used, and term entered use (especially for laptop displays) before the broad standardization 16:10 for widescreen computer displays.

Overall, several resolutions have been labeled as WXGA. 1280×720[16] provides perfectly square pixels at an aspect ratio of 16:9, while the additional pixels in 1280×768[17] and 1280×800 must be ignored to give the 16:9 ratio without vertical stretching of the image. 1360×768 and 1366×768 come very close to 16:9, displaying exactly square pixels if 1360×765 pixels of the display are used.

Recent widespread availability of 1280×800 pixel resolution LCDs for laptop monitors can be considered an OS driven evolution from the formerly popular 1024×768 screen size. In Microsoft Windows operating system specifically, the task bar when fit to the bottom of the screen occupies about 30 pixels, allowing a program window sized 1024x768 pixels to fit on screen without obstruction(800-768=32). Operating the Windows Sidebar in Windows Vista can use the remaining width of 256 pixels (1280-1024).

720p is a related HDTV video display resolution measuring 1280x720 pixels.

1440x900 resolution displays have also been found labeled as WXGA; however, the correct label is actually WSXGA or WXGA+.

XGA+ (1152x864)

Variants of XGA+

x (px)	y (px)	x:y	x·y (Mpx)	Origin
1152	900	5:4~4:3	0.983	Sun
1152	864	4:3	0.995	Apple

XGA+ stands for **Extended Graphics Array Plus** and is a computer display standard. XGA+ is often used on 17 inch desktop CRT monitors. XGA+ is usually understood to refer to the 1152x864 resolution with an aspect ratio of 4:3. As widescreen LCD are getting increasingly popular, this resolution is decreasing in use, but it is the native resolution of some 17 inch 4:3 LCD displays.

Historically, the resolution relates to the earlier standard of 1152x900 pixels, which was adopted by Sun Microsystems for the Sun-2 workstation in the early 1980s. This resolution is close to the maximum practical which, using one byte per pixel, can fit into a video memory or frame-buffer of one megabyte. However, its aspect ratio is 3.84:3 (1.28:1). When Apple Computer defined a standard resolution for 21-inch CRT monitors, intended for use as Two-Page Displays on the Macintosh II computer, Apple selected instead 1152x864, which is the highest 4:3 resolution below one million pixels.

XGA+ is the next step after XGA (1024x768), although it's not approved by any standard organizations. The next step with an aspect ratio of 4:3 is 1280x960 ("SXGA−") or SXGA+ (1400x1050).

WXGA+ (1440x900)

WXGA+ and **WSXGA** are non-standard terms referring to computer display resolutions. Usually they refer to a resolution of 1440x900, but occasionally manufacturers use other terms to refer to this resolution (for example, [18]). The Standard Panels Working Group refers to the 1440x900 resolution as WXGA(II).[19]

WSXGA and WXGA+ can be considered enhanced versions of WXGA with more pixels, or as widescreen variants of SXGA. The aspect ratios of each are 16:10 (widescreen).

WXGA+ (1440x900) resolution is common in 19" widescreen desktop monitors (a very small number of such monitors uses WSXGA+), and is also optional, although less common, in laptop LCDs, in sizes ranging from 12.1" to 17".

SXGA (1280×1024)

SXGA is an abbreviation for **Super Extended Graphics Array** referring to a standard monitor resolution of 1280×1024 pixels. This display resolution is the "next step" above the XGA resolution that IBM developed in 1990.

The 1280×1024 resolution is not the standard 4:3 aspect ratio, but 5:4 (1.25:1 instead of 1.333:1). A standard 4:3 monitor using this resolution will have rectangular rather than square pixels, meaning that unless the software compensates for this the picture will be distorted, causing circles to appear elliptical.

There is a less common 1280×960 resolution that preserves the common 4:3 aspect ratio. It is sometimes unofficially called **SXGA–** to avoid confusion with the "standard" SXGA. Elsewhere this 4:3 resolution was also called **UVGA** (*Ultra VGA*): Since both sides are doubled from VGA the term *Quad VGA* would be a systematic one, but it is hardly ever used, because its initialism QVGA is strongly associated with the alternate meaning *Quarter VGA* (320×240).

SXGA is the most common native resolution of 17" and 19" LCD monitors. An LCD monitor with SXGA native resolution will typically have a physical 5:4 aspect ratio, preserving a 1:1 pixel aspect ratio.

Sony manufactured a 17" CRT monitor with a 5:4 aspect ratio designed for this resolution. It was sold under the Apple brand name.

SXGA is also a popular resolution for cell phone cameras, such as the Motorola Razr and most Samsung and LG phones. Although being taken over by newer UXGA (2.0 megapixel) cameras, the 1.3 megapixel was the most common around 2007.

Any CRT that can run 1280×1024 can also run 1280×960, which has the standard 4:3 ratio. Displaying any 4:3 resolution on a 5:4 monitor, like a TFT with a native resolution of 1280×1024, will look stretched. But on a TFT, displaying any other resolution than the native one is not a good idea anyway, as the image needs to be interpolated to fit in the fixed grid display (and some TFT displays do not allow a user to disable this and use a letterbox format).

The 1280×1024 resolution became popular because at 24-bit color it fit well into 4 megabytes of video RAM. At the time, memory was extremely expensive. Using 1280×1024 at 24-bit color depth allowed using 3.75 MiB of video RAM, fitting nicely with VRAM chip sizes which were available at the time (4 MiB).

$$(1280 \times 1024)\text{ px} \times 8\text{ bit/px} \div 8\text{ bit/byte} \div 2^{20}\text{ byte/MiB} = 1.25\text{ MiB}$$

$$(1280 \times 1024)\text{ px} \times 24\text{ bit/px} \div 8\text{ bit/byte} \div 2^{20}\text{ byte/MiB} = 3.75\text{ MiB}$$

SXGA+ (1400×1050)

SXGA+ stands for **Super Extended Graphics Array Plus** and is a computer display standard. An SXGA+ display is commonly used on 14 inch or 15 inch laptop LCD screens with a resolution of 1400×1050 pixels. An SXGA+ display is used on a few 12 inch laptop screens such as the ThinkPad X60 and X61 (both only as tablet) as well as the Toshiba Portégé M200 and M400, but these are far less common. Dell offered a 14.1" SXGA+ screen on the many of the Dell Latitude "C" series laptops, such as the C640 and the C810. Sony also used SXGA+ in their Z1 series, but no longer produce them as wide screen has become more predominant.

There is a widescreen version of SXGA+ called WSXGA+ with a resolution of 1680×1050. This is a common native resolution of 19-22 inch wide-aspect LCD monitors, and is also available on many laptops.

It is the next common step in resolution after SXGA, although it is not approved by any organization. The most common resolution immediately above is called UXGA (sometimes also known as UGA) which has 1600×1200 pixels.

In desktop LCDs, SXGA+ is used on some low-end 20" monitors, whereas most of the 20" LCDs use UXGA (standard screen ratio), or WSXGA+ (widescreen ratio).

WSXGA+ (1680×1050)

WSXGA+ stands for **Widescreen Super Extended Graphics Array Plus** and is a computer display standard. A WSXGA+ display is commonly used on Widescreen 20", 21", and popular 22" LCD monitors from numerous manufacturers (and a very small number of 19" widescreen monitors), as well as widescreen 15.4" and 17" laptop LCD screens like the Thinkpad T61 and the Apple 15" MacBook Pro. The resolution is 1680×1050 pixels (1,764,000 pixels) and has a 16:10 aspect ratio.

WSXGA+ is the widescreen version of SXGA+, but it is not approved by any organization. The next highest resolution (for widescreen) after it is WUXGA, which is 1920×1200 pixels.

UXGA (1600×1200)

UXGA or **UGA** is an abbreviation for **Ultra Extended Graphics Array** referring to a standard monitor resolution of 1600×1200 pixels (totaling 1,920,000 pixels), which is exactly four times the default resolution of SVGA (800×600) (totaling 480,000 pixels). Dell Computer refers to the same resolution of 1,920,000 pixels as *UGA*. It is generally considered to be the next step above SXGA (1280×960 or 1280×1024), but some resolutions (such as the unnamed 1366×1024 and SXGA+ at 1400×1050) fit between the two.

UXGA has been the native resolution of many fullscreen monitors of 15" or more, including laptop LCDs such as the ones in ThinkPad A21p, A31p, T42p, and T43p; Dell Inspiron 8000/8100/8200; Panasonic Toughbook CF-51; and the original Alienware Area 51m. However, in more recent times, UXGA is not used in laptops at all but rather in desktop UXGA monitors that have been made in sizes of 20" and 21.3". Some 14" laptop LCDs with UXGA have also existed, but these were very rare.

There are two different widescreen cousins of UXGA, one called UWXGA with 1600×768 (750) and one called WUXGA with 1920×1200 resolution.

WUXGA (1920×1200)

WUXGA stands for **Widescreen Ultra Extended Graphics Array** and is a display resolution of 1920×1200 pixels (2,304,000 pixels) with a 16:10 screen aspect ratio.

It is a wide version of UXGA, and can be used for viewing high-definition television (HDTV) content, which uses a 16:9 aspect ratio and a 1920×1080 resolution. Because it shares the width of 1080p, it is sometimes referred to as 1200p, although this is technically incorrect.

The 16:10 aspect ratio (as opposed to the 16:9 used in widescreen televisions) was chosen because this aspect ratio is appropriate for displaying two full pages of text side by side.[20]

WUXGA resolution is equivalent to 2.304 megapixels. An 8-bit RGB WUXGA image has an uncompressed size of 6.912 MiB. This was the highest resolution that was ever commonly available in the computer display industry, but its use had been almost completely ended by 2010 as the computer industry moved to the 16:9 aspect ratio (i.e., 1920×1080 is the highest resolution available from most laptop and computer monitor manufacturers). This resolution is currently available in a few high-end LCD televisions and computer monitors, the latter of which are typically in the size range of approximately 23"–28" for desktop monitors, but has become almost completely unavailable on notebook monitors. A small number of 22" WUXGA desktop monitors exist (i.e., Lenovo L220x and Samsung T220P). WUXGA use predates the introduction of LCDs of that resolution. Most QXGA displays support 1920×1200 and widescreen CRTs such as the Sony GDM-FW900 and Hewlett Packard A7217A do as well.

The next lower resolution (for widescreen) before it is WSXGA+, which is 1680×1050 pixels (1,764,000 pixels, or 30.61% fewer than the WUXGA); the next higher resolution widescreen is an unnamed 2304×1440 resolution (supported by the above GDM-FW900 & A7217A) and then the more common WQXGA, which has 2560×1600 pixels (4,096,000 pixels, or 77.78% more than WUXGA).

There are two wider formats called UWXGA 1600×768 (25:12) and **UW-UXGA** that has 2560×1080 pixels, a 2.37:1 or 21⅓:9 or 64:27 aspect ratio, sometimes erroneously labeled 21:9.

Name	x(width)	y(height)	Mega-pixels	Aspect ratio	Percentage of difference in pixels						Typical sizes	Non-wide version	Note
					Wide XGA	WSXGA	WSXGA+	WUXGA	WQHD	WQXGA			
Wide XGA	1366	768	1.049	1.778		−24%	−68%	−120%	−251%	−290%	15"–19"	XGA	
WSXGA Wide XGA+	1440	900	1.296	1.6	+19%		−36%	−78%	−184%	−216%	15"–19"	XGA+	
WSXGA+	1680	1050	1.764	1.6	+41%	+27%		−31%	−109%	−132%	20"–22"	SXGA+	
WUXGA	1920	1200	2.304	1.6	+54%	+44%	+23%		−60%	−78%	23"–28"	UXGA	Displays 1920×1080 video with slight letterbox
WQHD	2560	1440	3.686	1.778	+72%	+65%	+52%	+38%		−11%	27"		
WQXGA	2560	1600	4.096	1.6	+74%	+68%	+57%	+44%	+10%		30"+	QXGA	Complements portrait UXGA

Quad XGA

Quad-Extended Graphics Array

Name	x (px)	y (px)	x:y	x·y (Mpx)
QWXGA	2048	1152	16:9	2.359
QXGA	2048	1536	4:3	3.145
WQXGA	2560	1600	16:10	4.096
QSXGA	2560	2048	5:4	5.242
WQSXGA	3200	2048	25:16	6.553
QUXGA	3200	2400	4:3	7.680
WQUXGA	3840	2400	16:10	9.216

The **QXGA**, or **Quad Extended Graphics Array**, display standard is a resolution standard in display technology. Some examples of LCD monitors that have pixel counts at these levels are the Dell 3008WFP, the Apple Cinema Display, the Apple iMac (27" 2009-present), the iPad (3rd generation), and the 2012 MacBook Pro. Many standard 21"/22" CRT monitors and some of the highest-end 19" CRTs also support this resolution.

QWXGA (2048×1152)

QWXGA (**Quad Wide Extended Graphics Array**) is a display resolution of 2048×1152 pixels with a 16:9 aspect ratio. A few LCD QWXGA monitors are available with 23 and 27 inch displays, such as the Acer B233HU (23") and B273HU (27"), the Dell SP2309W, and the Samsung 2343BWX.

QXGA (2048×1536)

QXGA (**Quad Extended Graphics Array**) is a display resolution of 2048×1536 pixels with a 4:3 aspect ratio. The name comes from it having four times as many pixels as an XGA display. Examples of LCDs with this resolution are the IBM T210 and the Eizo G33 and R31 screens, but in CRT monitors this resolution is much more common; some examples include the Sony F520,ViewSonic G225fB, NEC FP2141SB or Mitsubishi DP2070SB, Iiyama Vision Master Pro 514, and Dell and HP P1230. Of these monitors, none is still in production. A related display size is WQXGA, which is a wide screen version. CRTs offer a way to achieve QXGA cheaply. Models like the Mitsubishi Diamond Pro 2045U and IBM ThinkVision C220P retailed for around 200 USD, and even higher performance ones like the ViewSonic PerfectFlat P220fB remained under 500 USD. At one time, many off-lease P1230s could be found on eBay for under 150 USD. The LCDs with WQXGA or QXGA resolution typically cost 4 to 5 times more for the same resolution. IDTech manufactured a 15" QXGA IPS panel. NEC had sold laptops with QXGA screens in 2002-2005 for Japanese market.[21][22] The iPad (3rd generation) also has a QXGA display.[23]

WQXGA (2560×1600)

WQXGA (**Wide Quad Extended Graphics Array**) is a display resolution of 2560×1600 pixels with a 16:10 aspect ratio. The name comes from it being a wide version of QXGA and having four times as many pixels as an WXGA (1280×800) display.

To obtain a vertical refresh rate higher than 40 Hz, this resolution requires more bandwidth than a single link DVI supports and requires dual-link capable cables and devices. To avoid cable problems monitors are sometimes shipped with an appropriate dual link cable already plugged in. Many video cards support this resolution.

In 2010, WQXGA made its debut in a handful of home theater projectors targeted at the Constant Height Screen application market. Both Digital Projection Inc and projectiondesign released models based on a Texas Instrument DLP chip with a native WQXGA resolution, alleviating the need for an anamorphic lens to achieve 1:2.35 image projection.

One feature that is currently unique to the 30" WQXGA monitors is the ability to function as the centerpiece and main display of a three-monitor array of complementary aspect ratios, with two UXGA (1600×1200) 20" monitors turned vertically on either side. The resolutions are equal, and the size of the 1600 resolution edges (if the manufacturer is honest) is within a tenth of an inch (16" vs. 15.89999"), presenting a "picture window view" without the extreme lateral dimensions, small central panel, asymmetry, resolution differences, or dimensional difference of other three-monitor combinations. The resulting 4960×1600 composite image has a 3.1:1 aspect ratio. This also means one UXGA 20" monitor in portrait orientation can also be flanked by two 30" WQXGA monitors for a 6320×1600 composite image with a 11.85:3 (79:20, 3.95:1) aspect ratio. Some WQXGA medical displays (such as the Barco Coronis 4MP) can also be configured as two virtual 1200×1600 or 1280×1600 seamless displays by using both DVI ports at the same time.

Many manufacturers have 27"–30" models that are capable of WQXGA, albeit at a much higher price than lower resolution monitors of the same size. Several mainstream WQXGA monitors are available with 30 inch displays, such as the Dell UltraSharp 3007WFP-HC, Dell UltraSharp 3008WFP and Dell UltraSharp 3011, the Hewlett-Packard LP3065, the Gateway XHD3000, LG W3000H, and the Samsung 305T. Specialist manufacturers like Eizo, Planar Systems, Barco (LC-3001), and possibly others offer similar models.

QSXGA (2560×2048)

QSXGA (**Quad Super Extended Graphics Array**) is a display resolution of 2560×2048 pixels with a 5:4 aspect ratio. Grayscale monitors with a 2560×2048 resolution, primarily for medical use, are available from Planar Systems (Dome E5), Eizo (Radiforce G51), Barco (Nio 5,MP), WIDE (IF2105MP), IDTech (IAQS80F), and possibly others.

Recent medical displays such as Barco Coronis Fusion 10MP or NDS Dome S10 have native panel resolution of 4096×2560. These are driven by two dual link DVI or displayports. They can be considered to be two seamless virtual QSXGA displays as they have to be driven simultaneously by both dual link DVI or displayport since one dual link DVI or displayport cannot single-handedly display 10 megapixels.

A similar resolution of 2560×1920 (4:3) was supported by a small number of CRT displays via VGA such as the Viewsonic P225f when paired with the right graphics card.

WQSXGA (3200×2048)

WQSXGA (**Wide Quad Super Extended Graphics Array**) describes a display standard that can support a resolution up to 3200×2048 pixels, assuming a 1.56:1 (25:16) aspect ratio. The Coronis Fusion 6MP DL by Barco supports 3280×2048 (approx. 16:10).

QUXGA (3200×2400)

QUXGA (**Quad Ultra Extended Graphics Array**) describes a display standard that can support a resolution up to 3200×2400 pixels, assuming a 4:3 aspect ratio.

WQUXGA (3840×2400)

WQUXGA (**Wide Quad Ultra Extended Graphics Array**) describes a display standard that supports a resolution of 3840×2400 pixels, which provides a 16:10 aspect ratio. This resolution is exactly four times 1920×1200 (in pixels).

The IBM T221 display

WQUXGA is the maximum resolution supported by DisplayPort 1.2, though actually displaying such a resolution on a device with DisplayPort 1.2 is dependent on the graphics system in much the same way devices with VGA connectors do not necessarily maximize that standard's highest possible resolution.

In June 2001, WQUXGA was introduced in the IBM T220 LCD monitor using a LCD panel built by IDTech. LCD displays that support WQUXGA resolution include: IBM T220, IBM T221 (models DG1, DG3, DG4, DG5), Iiyama AQU5611DTBK, ViewSonic VP2290,[24] ADTX MD22292B, and IDTech MD22292 (models B0, B1, B2, B5, C0, C2). IDTech was the original equipment manufacturer which sold these monitors to ADTX, IBM, Iiyama, and ViewSonic.[25]

Most display cards with a DVI connector are capable of supporting the 3840×2400 resolution. However, the maximum refresh rate will be limited by the number of DVI links which are connected to the monitor. 1, 2, or 4 DVI connectors are used to drive the monitor using various tile configurations. Only the IBM T221-DG5 and IDTech MD22292B5 support the use of dual-link DVI ports through an external converter box.

Many systems using these monitors use at least 2 DVI connectors to send video to the monitor. These DVI connectors can be from the same graphics card, different graphics cards, or even different computers. Motion across the tile boundary(ies) can show tearing if the DVI links are not synchronized. The display panel can be updated at a speed between 0 Hz and 41 Hz (48 Hz for the IBM T221-DG5 and IDTech MD22292B5). The refresh rate of the video signal can be higher than 41 Hz (or 48 Hz) but the monitor will not update the display any faster even if graphics card(s) do so.

There was one series of WQUXGA displays in the consumer marketplace, but it was discontinued in Q2 of 2005. That series of displays had prices which were well above even the higher end displays used by graphic professionals. In addition, the lower refresh rates, 41 Hz and 48 Hz, made them less attractive for many applications. None of the WQUXGA monitors (IBM, ViewSonic, Iiyama, ADTX) are in production anymore.

Hyper XGA

Hyper-Extended Graphics Array

Name	x (px)	y (px)	x:y	x·y (Mpx)
HXGA	4096	3072	4:3	12.582
WHXGA	5120	3200	16:10	16.384
HSXGA	5120	4096	5:4	20.971
WHSXGA	6400	4096	25:16	26.214
HUXGA	6400	4800	4:3	30.720
WHUXGA	7680	4800	16:10	36.864

The **HXGA** display standard and its derivatives are a standard in display technology. As of 2012, there is no monitor that displays at these levels but several digital cameras can record such images. An example can be found in HIPerSpace [26] of a case where multiple WQXGA displays must be stacked to exceed HXGA resolution.

HXGA (4096×3072)

HXGA an abbreviation for **Hex**[adecatuple] Extended **G**raphics **A**rray is a display standard that can support a resolution of 4096×3072 pixels (or 3200 pixels) with a 4:3 aspect ratio. The name comes from it having sixteen (*hexadecatuple*) times as many pixels as an XGA display. A related display size is WHXGA, which is a wide screen version.

WHXGA (5120×3200)

WHXGA an abbreviation for **W**ide **H**ex[adecatuple] Extended **G**raphics **A**rray is a display standard that can support a resolution of roughly 5120x3200 pixels with a 16:10 aspect ratio. The name comes from it being a wide version of HXGA, which has sixteen (*hexadecatuple*) times as many pixels as an XGA display. It would require four WQXGA devices to display at this resolution. A resolution of 5120×3072 should, in theory, also qualify as WHXGA, if such a display were to be made.

HSXGA (5120×4096)

HSXGA, an abbreviation for **H**ex[adecatuple] **S**uper Extended **G**raphics **A**rray, is a display standard that can support a resolution of roughly 5120×4096 pixels with a 5:4 aspect ratio. The name comes from it having sixteen (*hexadecatuple*) times as many pixels as an SXGA display.

WHSXGA (6400×4096)

WHSXGA, an abbreviation for **W**ide **H**ex[adecatuple] **S**uper Extended **G**raphics **A**rray, is a display standard that can support a resolution up to 6400×4096 pixels, assuming a 1.56:1 (25:16) aspect ratio. The name comes from it having sixteen (*hexadecatuple*) times as many pixels as an WSXGA display.

HUXGA (6400×4800)

HUXGA, an abbreviation for **Hex**[adecatuple] **U**ltra E**x**tended **G**raphics **A**rray, is a display standard that can support a resolution of roughly 6400×4800 pixels with a 4:3 aspect ratio. The name comes from it having sixteen (*hexadecatuple*) times as many pixels as an UXGA display.

WHUXGA (7680×4800)

WHUXGA an abbreviation for **W**ide **H**ex[adecatuple] **U**ltra E**x**tended **G**raphics **A**rray, is a display standard that can support a resolution up to 7680×4800 pixels, assuming a 8:5 (16:10) aspect ratio. The name comes from it having sixteen (*hexadecatuple*) times as many pixels as a WUXGA display. A WHUXGA image consists of 36,864,000 pixels (approximately 37 megapixels). UHDTV requires a display of similar resolution (7680×4320) for properly displaying UHDTV content, which is 16 times the pixel count of the 1080p ATSC HDTV video standard.

High-Definition

High-Definition

Name	x (px)	y (px)	x:y	x·y (Mpx)
nHD	640	360	16:9	0.230
qHD	960	540	16:9	0.518
HD	1280	720	16:9	0.921
HD+	1600	900	16:9	1.44
FHD	1920	1080	16:9	2.073
QHD	2560	1440	16:9	3.686
QFHD	3840	2160	16:9	8.294
UHD	7680	4320	16:9	33.178

nHD (640×360)

nHD is a display resolution of 640×360 pixels, which is exactly one **n**inth of a Full HD (1080p) frame and one quarter of a HD (720p) frame. 2×2 nHD frames will form one 720p frame and 3×3 nHD frames will form one 1080p frame.

One drawback of this resolution is that the vertical resolution is not an even multiple of 16, which is a common macroblock size for video codecs. Video frames encoded with 16×16 pixel macroblocks would be padded to 640×368 and the added pixels would be cropped away at playback. The same is true for qHD and 1080p but the relative amount of padding is more for lower resolutions such as nHD.

To avoid storing padding data, some people prefer to encode video at 624×352. When such video streams are either encoded from HD frames or played back on HD displays in full screen mode (either 720p or 1080p) they are scaled by non-integer scale factors. True nHD frames on the other hand has integer scale factors.

qHD (960×540)

qHD is a display resolution of 960×540 pixels, which is exactly one **q**uarter of a Full HD (1080p) frame, in a 16:9 aspect ratio.

Similar to DVGA, this resolution became popular for high-end smartphone displays in early 2011. Mobile phones including the HTC Sensation, HTC Evo 3D, Motorola Droid RAZR, and Motorola Atrix 4G have displays with the qHD resolution, as does the PlayStation Vita portable game system.

HD (1280×720)

The **HD** resolution of 1280×720 pixels stems from high-definition television (HDTV), where it originally used 60 frames per second. With its 16:9 aspect ratio it is exactly 2 times the width and 1 $^1/_2$ times the height of 4:3 VGA, which shares its aspect ratio and 480 line count with NTSC. HD therefore has exactly 3 times as many pixels as VGA.

This resolution is sometimes referred to as **720p**, although the *p* (which stands for progressive scan and is important for transmission formats) is irrelevant for labeling digital display resolutions.

Few screens have been built that actually use this resolution natively, most employ 16:9 panels with 768 lines instead (WXGA), which resulted in odd numbers of pixels per line, i.e. 1365 $^1/_3$ are rounded to 1360, 1364, 1366 or even 1376, the next multiple of 16. All of these resolutions are in scope of the "HD ready" label. The HTC Rezound smartphone has this kind of display.

FHD (1920×1080)

The **FHD** or **Full HD** resolution of 1920×1080 pixels in a 16:9 aspect ratio was developed as an HDTV transmission and storage format using interlacing and then have bandwidth demands very similar to those of 720p, therefore their pixel counts are roughly in a 2:1 ratio, 9:4 exactly. FHD is 3 times the width and 2 $^1/_4$ times the height of 4:3 VGA.

Due to its origins described above, this resolution is sometimes referred to as **1080i**, wherein the **i** stands for "interlaced". Since there are also progressive signals with the same frame rate, but half the field rate, or less, it sometimes is also called **1080p**.

Since most video codecs use 16×16 pixel macro blocks there is often an excess 8 lines encoded, for 16 times 68 equals 1088.

WQHD (2560×1440)

WQHD (Wide Quad High Definition) is a display resolution of 2560×1440 pixels in a 16:9 aspect ratio. It is four times the pixel resolution of the 720p HDTV video standard, hence the name.

This resolution was under consideration by the ATSC in the late 1990s to become the standard HDTV format, because it is exactly 4 times the width and 3 times the height of VGA, which has the same amount of lines as NTSC signals at the SDTV 4:3 aspect ratio. Pragmatic technical constraints made them choose the now well-known 16:9 formats with twice (HD) and thrice (FHD) the VGA width instead.

In autumn 2006, Chi Mei Optoelectronics (CMO) announced a 47" 1440 LCD panel to be released in Q2 2007;[27] the panel was planned to finally debut at FPD International 2008 in a form on autostereoscopic 3D display.[28]

Some examples of LCD monitors that have pixel counts at these levels are the Dell UltraSharp U2711,[29] U2713HM[30] and XPS One 27", Samsung S27A850D, S27A970, and S27B970, HP ZR2740W, ViewSonic VP2770-LED, Nixeus NX-VUE27, NEC MultiSync PA271W,[31] and the Apple LED Cinema Display, Thunderbolt Display, and 27" iMac.[32]

QFHD (3840×2160)

QFHD (Quad Full High Definition), also known as **4K** or **4K UHDTV**, is a display resolution of 3840×2160 pixels in a 16:9 aspect ratio.[33] In 2012 ITU drafts, this resolution is part of the UHDTV standard.[34] It is four times the resolution of the 1080p HDTV video standard, hence the name (Quad meaning 4). HDMI 1.4 supports QFHD.[35]

In early 2008, Samsung revealed a proof-of-concept 82-inch LCD TV set capable of this resolution[36] and LG has demonstrated an 84-inch display.[37]

CMI has built a 27.84" 158 PPI QFHD IPS panel for medical displays since November 2010.[38] Optik View has two versions of 56" QFHD monitors. DC801 has 2 Dual Link DVI input; DC802 has 4 different versions: 4 single link DVI, 4 HDMI, 4 DisplayPort and/or 4 3G-SDI inputs. All version can deliver a resolution of 3840x2160.[39] Eyevis produces a 56" LCD named EYELCD 56 QHD HD while Toshiba makes the P56QHD and in October 2011 released the REGZA 55x3,[40] which is claimed to be the First QFHD glasses-free 3D TV, Mitsubishi Electric the 56P-QF60LCU, and Sony the SRM-L560, all which can deliver a resolution of 3840×2160.[41] Landmark has also produced a 56" QFHD monitor, the M5600.[42]

UHD (7680×4320)

UHD (Ultra High Definition), also known as **8K** or **8K UHDTV**, is a proposed display standard of 7680×4320 pixels (16 times the resolution of FHD) in the same 16:9 aspect ratio.[34] It is advocated by NHK Science & Technology Research Laboratories.

References

[1] Kwon, Jang Yeon; Jung, Ji Sim; Park, Kyung Bae; Kim, Jong Man; Lim, Hyuck; Lee, Sang Yoon; Kim, Jong Min; Noguchi, Takashi et al. (2006), "2.2 inch qqVGA AMOLED Drove by Ultra Low Temperature Poly Silicon (ULTPS) TFT Direct Fabricated Below 200°C", *SID 2006 Digest* **37** (2): 1358–1361, doi:10.1889/1.2433233

[2] "QVGA" (http://www.topbits.com/qvga.html). . Retrieved February 10, 2010.

[3] Shin, Min-Seok; Choi, Jung-Whan; Kim, Yong-Jae; Kim, Kyong-Rok; Lee, Inhwan; Kwon, Oh-Kyong (2007). "Accurate Power Estimation of LCD Panels for Notebook Design of Low-Cost 2.2-inch qVGA LTPS TFT-LCD Panel". *SID 2007 Digest* **38** (1): 260–263

[4] "Optoma DLP Pico projector 'coming soon' to US" (http://www.engadget.com/2008/11/24/optoma-dlc-pico-projector-coming-soon-to-us/). engadget. 2008-11-24. . Retrieved 2008-11-24.

[5] "Supporting Multiple Screens" (http://developer.android.com/guide/practices/screens_support.html). . Retrieved 2011-02-04.

[6] Ken Polsson. "Chronology of IBM Personal Computers" (http://www.islandnet.com/~KPOLSSON/ibmpc/ibm1987.htm). . Retrieved 2010-11-18.

[7] "New resolutions for Microsoft Smartphone" (http://msmobiles.com/news.php/1541.html). ***.Dead Link***

[8] http://www.nvidia.com/docs/IO/55043/NVIDIA_Tegra_FAQ_External.pdf

[9] "Smart Computing Encyclopedia Entry - Ultra Video Graphics Array (UVGA)" (http://www.smartcomputing.com/editorial/dictionary/detail.asp?guid=&searchtype=&DicID=19364&RefType=Encyclopedia). Smartcomputing.com. 2007-01-16. . Retrieved 2012-02-21.

[10] Vipul Verma. "Same monitor yet better viewing" (http://www.tribuneindia.com/2001/20011029/login/main3.htm). www.tribuneindia.com. . Retrieved 2008-03-26.

[11] Hitachi plasma TVs, 1366×768 as WXGA (http://www.hitachi.ca/supportingdocs/en/forhome/plasma_tvs/plasma_chart_full.pdf)PDF

[12] TV Panels Standard VESA TV Panels Standard (http://www.vesa.org/Public/Panel Standards/TVpnlV1.pdf)PDF

[13] Microsoft PowerPoint - VESA Asia presentations (http://www.vesa.org/press/AsiaTourOct05.pdf)PDF—slide 21

[14] Dell laptop displays, 1280×800 as WXGA (http://www.dell.com/content/learnmore/learnmore.aspx?c=us&cs=04&l=en&s=bsd&~id=screen&~line=notebooks&~mode=popup&~model=d800&~series=latit&ref=CFG)

[15] static Lenovo laptop displays, 1280×800 as WXGA (http://shop.lenovo.com/SEUILibrary/controller/e/web/LenovoPortal/en_US/systemconfig.runtime.workflow:GetMoreInfo?fCode=/merchandising/US/specialoffers/popups/help_me_decide/Popup_helpme_display.html)

[16] Acer projector, 1280×720 as WXGA (http://www.ncix.com/products/index.php?sku=22206&vpn=EY.J4401.007&manufacture=Acer)

[17] Planar 17" LCD monitor, 1280×768 as WXGA (http://www.pricegrabber.ca/search_getprod.php/masterid=29941499//)

[18] http://www.renesas.com/fmwk.jsp?cnt=press_release20050912.htm&fp=/company_info/news_and_events/press_releases

[19] "Standard Panels Working Group" (http://www.spwg.org/specifications.htm). .

[20] "Introduction—Monitor Technology Guide" (http://web.archive.org/web/20070315085244/http://www.necdisplay.com/support/css/monitortechguide/index05.htm). necdisplay.com. Archived from the original (http://www.necdisplay.com/support/css/monitortechguide/

index05.htm) on 2007-03-15. .
[21] N E C : : 2 0 0 2 07 01 -1 (http://www.nec.co.jp/press/ja/0207/0101.html)
[22] N E C : : 2 0 0 5 01 19 -1 (http://www.nec.co.jp/press/ja/0501/1901.html)
[23] "Apple unveils new iPad with high-definition screen" (http://www.bbc.co.uk/news/technology-17292424). *BBC News* (BBC). March 7, 2012. . Retrieved March 7, 2012.
[24] ViewSonic: Company Info: Press Center: Press Releases (http://www.viewsonic.com/companyinfo/pressrelease_detail.cfm?key_press_release=155)
[25] About Purchase of the Ultra High-Resolution and Ultra High-Density LCD Monitor (http://www.idtech.co.jp/en/920LCD/how2buy.html)
[26] http://vis.ucsd.edu/mediawiki/index.php/Research_Projects:_HIPerSpace
[27] "CMO to ship 47-inch Quad HD — 1440 — LCD in 2007" (http://www.engadgethd.com/2006/10/17/cmo-to-ship-47-inch-quad-hd-1440-lcd-in-2007). Chi Mei Optoelectronics. 2006-10-17. . Retrieved 2008-07-06.
[28] "CMO showcases latest "green" and "innovative" LCD panel" (http://www.cmo.com.tw/opencms/cmo/modules/news/MCNews/mcnews_0111.html?__locale=en). Chi Mei Optoelectronics. 2008-10-24. . Retrieved 2008-10-26.
[29] "Dell UltraSharp U2711 69cm (27") Monitor with PremierColor Details" (http://accessories.us.dell.com/sna/products/Displays/productdetail.aspx?c=us&l=en&s=bsd&cs=04&sku=224-8284). Dell. . Retrieved 2010-08-28.
[30] "Dell™ UltraSharp U2713HM 27" Monitor with LED Details" (http://accessories.ap.dell.com/sna/productdetail.aspx?c=au&cs=audhs1&l=en&s=dhs&sku=210-40773). Dell. . Retrieved 2012-08-09.
[31] "NEC MultiSync PA271W" (http://reviews.cnet.com/lcd-monitors/nec-multisync-pa271w/4505-3174_7-34108768.html). cnet. 2010-07-07. . Retrieved 2010-08-28.
[32] "iMac line updated with 16:9 displays, quad-core Core i5 / i7 model" (http://www.engadget.com/2009/10/20/imac-line-updated-with-16-9-displays-quad-core-core-i5-model). engadget. 2009-10-20. . Retrieved 2010-08-28.
[33] Panasonic press release for 20.4-inch "4k2k" display (http://www2.panasonic.com/webapp/wcs/stores/servlet/prModelDetail?&itemId=664010)
[34] ITU video Ultra High Definition Television: Threshold of a new age (http://www.youtube.com/watch?v=hT2XluvAjwQ)
[35] "4K Support" (http://www.hdmi.org/manufacturer/hdmi_1_4/4K.aspx). http://www.hdmi.org. .
[36] Concept Samsung 82-Inch LCD World's Largest Ultra High-Definition (http://gizmodo.com/342997/concept-samsung-82+inch-lcd-worlds-largest-ultra-high+definition)
[37] LG Shows Off 84-Inch 3DTV With 3840×2160 Resolution (http://gizmodo.com/5547081/lg-shows-off-84+inch-3dtv-with-3840-x-2160-resolution)
[38] "27.8"(R278D1)" (http://www.chimei-innolux.com/opencms/cmo/products/medical_display/products_medical_R278D1.html?__locale=en). Chimei Innolux Corporation. . Retrieved 27 December 2010.
[39] "Optik View DC801, DC802 QFHD" (http://www.optikview.com/?p=228). The Linden Group Corp. . Retrieved 19 July 2010.
[40] "Toshiba's REGZA 55x3 announced as world's first 4K2K TV with glasses-free 3D" (http://www.engadget.com/photos/toshibas-regza-55x3-announced-as-worlds-first-4k2k-tv-with-glasses-free-3d/). http://www.engadget.com. .
[41] "EYE-LCD 5600 QHD" (http://www.eyevis.de/index.php?article_id=50&clang=1). eyevis GmbH. . Retrieved 5 July 2010.
[42] (http://www.halliburton.com/ps/default.aspx?pageid=1749&navid=964&prodid=PRN::JO84B3LPT)

Java_Platform,_Micro_Edition

Java Platform, Micro Edition, or **Java ME**, is a Java platform designed for embedded systems (mobile devices are one kind of such systems). Target devices range from industrial controls to mobile phones (especially feature phones) and set-top boxes. Java ME was formerly known as **Java 2 Platform, Micro Edition** (**J2ME**).

Java ME was designed by Sun Microsystems, acquired by Oracle Corporation in 2010; the platform replaced a similar technology, PersonalJava. Originally developed under the Java Community Process as JSR 68, the different flavors of Java ME have evolved in separate JSRs. Sun provides a reference implementation of the specification, but has tended not to provide free binary implementations of its Java ME runtime environment for mobile devices, rather relying on third parties to provide their own.

As of 22 December 2006, the Java ME source code is licensed under the GNU General Public License, and is released under the project name phoneME.

As of 2008, all Java ME platforms are currently restricted to JRE 1.3 features and use that version of the class file format (internally known as version 47.0). Should Oracle ever declare a new round of Java ME configuration versions that support the later class file formats and language features, such as those corresponding JRE 1.5 or 1.6 (notably, generics), it will entail extra work on the part of all platform vendors to update their JREs.

Java ME devices implement a *profile*. The most common of these are the Mobile Information Device Profile aimed at mobile devices, such as cell phones, and the Personal Profile aimed at consumer products and embedded devices like set-top boxes and PDAs. Profiles are subsets of *configurations*, of which there are currently two: the Connected Limited Device Configuration (CLDC) and the Connected Device Configuration (CDC).[1]

There are more than 2.1 billion Java ME enabled mobile phones and PDAs.[2] Although it not used on some of today's newest mobile platforms (e.g. iPhone, Windows Phone 7, BlackBerry 10, Android), it continues to be very popular in sub $200 devices such as Nokia's Series 40. It is also used on new Bada operating system and on Symbian OS along with native software. Also, there is an implementation for Android available for separate download.[3]

Connected Limited Device Configuration

The Connected Limited Device Configuration (CLDC) contains a strict subset of the Java-class libraries, and is the minimum amount needed for a Java virtual machine to operate. CLDC is basically used for classifying myriad devices into a fixed configuration.

A configuration provides the most basic set of libraries and virtual-machine features that must be present in each implementation of a J2ME environment. When coupled with one or more profiles, the Connected Limited Device Configuration gives developers a solid Java platform for creating applications for consumer and embedded devices. The configuration is designed for devices with 160KB to 512KB total memory, which has a minimum of 160KB of ROM and 32KB of RAM available for the Java platform.

Mobile Information Device Profile

Designed for mobile phones, the Mobile Information Device Profile includes a GUI, and a data storage API, and MIDP 2.0 includes a basic 2D gaming API. Applications written for this profile are called MIDlets. Almost all new cell phones come with a MIDP implementation, and it is now the de facto standard for downloadable cell phone games. However, many cellphones can run only those MIDlets that have been approved by the carrier, especially in North America.

JSR 271: Mobile Information Device Profile 3 (Final release on 09 Dec, 2009) specified the 3rd generation Mobile Information Device Profile (MIDP3), expanding upon the functionality in all areas as well as improving interoperability across devices. A key design goal of MIDP3 is backward compatibility with MIDP2 content.

Information Module Profile

The Information Module Profile (IMP) is a profile for embedded, "headless" devices such as vending machines, industrial embedded applications, security systems, and similar devices with either simple or no display and with some limited network connectivity.

Originally introduced by Siemens Mobile and Nokia as JSR-195, IMP 1.0 is a strict subset of MIDP 1.0 except that it doesn't include user interface APIs — in other words, it doesn't include support for the Java package `javax.microedition.lcdui`. JSR-228, also known as IMP-NG, is IMP's next generation that is based on MIDP 2.0, leveraging MIDP 2.0's new security and networking types and APIs, and other APIs such as `PushRegistry` and `platformRequest()`, but again it doesn't include UI APIs, nor the game.

Connected Device Configuration

The Connected Device Configuration is a subset of Java SE, containing almost all the libraries that are not GUI related. It is richer than CLDC.

Foundation Profile

The Foundation Profile is a Java ME Connected Device Configuration (CDC) profile. This profile is intended to be used by devices requiring a complete implementation of the Java virtual machine up to and including the entire Java Platform, Standard Edition API. Typical implementations will use some subset of that API set depending on the additional profiles supported. This specification was developed under the Java Community Process.

Personal Basis Profile

The Personal Basis Profile extends the Foundation Profile to include lightweight GUI support in the form of an AWT subset. This is the platform that BD-J is built upon.

Implementations

Sun provides a reference implementation of these configurations and profiles for MIDP and CDC. Starting with the JavaME 3.0 SDK, a NetBeans-based IDE will support them in a single IDE.

In contrast to the numerous binary implementations of the Java Platform built by Sun for servers and workstations, Sun does not provide any binaries for the platforms of Java ME targets with the exception of an MIDP 1.0 JRE (JVM) for Palm OS.[4] Sun provides no J2ME JRE for the Microsoft Windows Mobile (Pocket PC) based devices, despite an open-letter campaign to Sun to release a rumored internal implementation of PersonalJava known by the code name "Captain America".[5] Third party implementations like JBlend and JBed are widely used by Windows Mobile vendors like HTC and Samsung.

Operating systems targeting Java ME have been implemented by DoCoMo in the form of DoJa, and by SavaJe as SavaJe OS. The latter company was purchased by Sun in April 2007 and now forms the basis of Sun's JavaFX Mobile. The company IS2T provides Java ME virtual machine (MicroJvm), for any RTOS and even with no-RTOS then qualified as baremetal. When baremetal, the virtual machine is the OS/RTOS: the device boots in Java.[6]

MicroEmulator [7] provides an open source (LGPL) implementation of MIDP emulator. This is a Java Applet based emulator and can be embedded in web pages.

The open-source Mika VM aims to implement JavaME CDC/FP, but is not certified as such (certified implementations are required to charge royalties, which is impractical for an open-source project). Consequently devices which use this implementation are not allowed to claim JavaME CDC compatibility.

JSRs (Java Specification Requests [8])

Foundation

JSR #	Name	Description
68 [9]	J2ME Platform Specification	
30 [10]	CLDC 1.x	
37 [11]	MIDP 1.0	
118 [12]	MIDP 2.x	
139 [13]	CLDC 1.1	
271 [14]	MIDP 3.0	

Future

JSR #	Name	Description
297 [15]	Mobile 3D Graphics API (M3G) 2.0	

Main extensions

JSR #	Name	Description
75 [16]	File Connection and PIM	File system, contacts, calendar, to-do
82 [17]	Bluetooth	
120 [18]	Wireless Messaging API (WMA)	
135 [19]	Mobile Media API (MMAPI)	Audio, video, multimedia
172 [20]	Web Services	
177 [21]	Security and Trust Services	
179 [22]	Location API	
180 [23]	SIP API	
184 [24]	Mobile 3D Graphics	High level 3D graphics
185 [25]	Java Technology for the Wireless Industry (JTWI)	General
205 [26]	Wireless Messaging 2.0 (WMA)	
211 [27]	Content Handler API	
226 [28]	SVG 1.0	
229 [29]	Payment API	
234 [30]	Advanced Multimedia Supplements (AMMS)	MMAPI extensions
238 [31]	Mobile Internationalization API	

239 [32]	Java Bindings for the OpenGL ES API	
248 [33]	Mobile Service Architecture	General
256 [34]	Mobile Sensor API	
287 [35]	SVG 2.0	

ESR

The ESR consortium is devoted to Standards for embedded Java. Especially cost effective Standards. Typical applications domains are industrial control, machine-to-machine, medical, e-metering, home automation, consumer, human-to-machine-interface, ...

ESR #	Name	Description
001 [36]	B-ON (Beyond CLDC)	B-ON serves as a very robust foundation for implementing embedded Java software. It specifies a reliable initialization phase of the Java device, and 3 kind of objects: immutable, immortal and regular (mortal) objects.
002 [36]	MicroUI	MicroUI defines an enhanced architecture to enable an open, third-party, application development environment for embedded HMI devices. Such devices typically have some form of display, some input sensors and potentially some sound rendering capabilities. This specification spans a potentially wide set of devices.
011 [36]	MWT	MWT defines three distinct roles: Widget Designers, Look and Feel Designers and Application Designers. MWT allows same binary HMI application to run the same on all devices that provide a compliant MWT framework (embedded devices, cellphones, setopbox-TV, PC, etc...) allowing for true ubiquity of applications across product lines (ME, SE, EE).
015 [36]	ECLASSPATH	ECLASSPATH unifies CLCD, CDC, Foundation, SE, EE execution environments with a set of around 300 classes API. Compiling against CLDC1.1/ECLASSPATH makes binary code portable across all Java execution environments.

Notes

[1] Java ME Technology (http://java.sun.com/javame/technology/)
[2] About Java (http://www.java.com/en/about/)
[3] App Runner (http://www.netmite.com/android/)
[4] MIDP for Palm OS 1.0: Developing Java Applications for Palm OS Devices (http://developers.sun.com/mobility/midp/articles/palm/) January 2002
[5] CDC and Personal Profile - Open letter to SUN to produce a Personal Java JRE for Pocket PC (http://forum.java.sun.com/thread.jspa?threadID=408223) 2003
[6] IS2T (http://www.is2t.com/)
[7] http://www.microemu.org/
[8] http://jcp.org/en/jsr/overview
[9] http://www.jcp.org/en/jsr/detail?id=68
[10] http://www.jcp.org/en/jsr/detail?id=30
[11] http://www.jcp.org/en/jsr/detail?id=37
[12] http://www.jcp.org/en/jsr/detail?id=118
[13] http://www.jcp.org/en/jsr/detail?id=139
[14] http://www.jcp.org/en/jsr/detail?id=271
[15] http://www.jcp.org/en/jsr/detail?id=297
[16] http://www.jcp.org/en/jsr/detail?id=75
[17] http://www.jcp.org/en/jsr/detail?id=82
[18] http://www.jcp.org/en/jsr/detail?id=120
[19] http://www.jcp.org/en/jsr/detail?id=135
[20] http://www.jcp.org/en/jsr/detail?id=172
[21] http://www.jcp.org/en/jsr/detail?id=177
[22] http://www.jcp.org/en/jsr/detail?id=179

[23] http://www.jcp.org/en/jsr/detail?id=180
[24] http://www.jcp.org/en/jsr/detail?id=184
[25] http://www.jcp.org/en/jsr/detail?id=185
[26] http://www.jcp.org/en/jsr/detail?id=205
[27] http://www.jcp.org/en/jsr/detail?id=211
[28] http://www.jcp.org/en/jsr/detail?id=226
[29] http://www.jcp.org/en/jsr/detail?id=229
[30] http://www.jcp.org/en/jsr/detail?id=234
[31] http://www.jcp.org/en/jsr/detail?id=238
[32] http://www.jcp.org/en/jsr/detail?id=239
[33] http://www.jcp.org/en/jsr/detail?id=248
[34] http://www.jcp.org/en/jsr/detail?id=256
[35] http://www.jcp.org/en/jsr/detail?id=287
[36] http://www.e-s-r.net/en/esrlist.php

- JSR 232: Mobile Operational Management (http://www.jcp.org/en/jsr/detail?id=232) An advanced OSGi technology based platform for mobile computing
- JSR 291: Dynamic Component Support for Java SE (http://www.jcp.org/en/jsr/detail?id=291) - Symmetric programming model for Java SE to Java ME JSR 232

Bibliography

- Hayun, Roy Ben (March 30, 2009). *Java ME on Symbian OS: Inside the Smartphone Model* (http://eu.wiley.com/WileyCDA/WileyTitle/productCd-0470743182.html) (1st ed.). Wiley. p. 482. ISBN 0-470-74318-2.
- Knudsen, Jonathan (January 8, 2008). *Kicking Butt with MIDP and MSA: Creating Great Mobile Applications* (http://www.informit.com/store/product.aspx?isbn=9780321463425) (1st ed.). Prentice Hall. p. 432. ISBN 0-321-46342-0.
- Li, Sing; Knudsen, Jonathan (April 25, 2005). *Beginning J2ME: From Novice to Professional* (http://www.apress.com/book/view/9781590594797) (3rd ed.). Apress. p. 480. ISBN 1-59059-479-7.

External links

- Sun Developer Network, Java ME (http://java.sun.com/javame/index.jsp)
- J2ME Game Developer Network (http://www.j2megame.org/)
- Nokia's Developer Hub Java pages (http://www.forum.nokia.com/java)
- Nokia S60 Java Runtime blogs (http://blogs.s60.com/java)
- Sony Ericsson Developer World (http://developer.sonyericsson.com/)
- Motorola Developer Network (http://developer.motorola.com/)
- J2ME Authoring Tool LMA Users Network (http://hotlavasoftware.com/)
- Samsung Mobile Developer's site (http://developers.samsungmobile.com/)
- Sprint Application Developer's Website (http://developer.sprint.com/)
- Performance database of Java ME compatible devices (http://www.jbenchmark.com/)
- IS2T J2ME platforms for embedded systems (http://www.is2t.com/)
- Book - Mobile Phone Programming using Java ME (J2ME) (http://www.skjapp.com/javame-j2me)

Article Sources and Contributors

Nokia_C2-02 *Source*: http://en.wikipedia.org/w/index.php?title=Nokia_C2-02 *Contributors*: ItsMeOrYou, Mrjessierama, 2 anonymous edits

Nokia *Source*: http://en.wikipedia.org/w/index.php?title=Nokia *Contributors*: -Majestic-, 007zoo, 130.236.221.xxx, 159753, 16@r, 1exec1, 205ywmpq, 28421u2232nfenfcenc, A bit iffy, A box of sticks, A. B., A333, AB, ABF, AEMoreira042281, Abdul raja, Abnyc81, Academic Challenger, Accurate Nuanced Clear, Acdx, Acela Express, Adamrush, Adaobi, Adashiel, Adraeus, Aesopos, Aeusoes1, Agusm266, Ahazred8, Ahmedcena, Ahoerstemeier, Aijoovai, Aintneo, Alansohn, Aldis90, Alepik, Alex43223, Alexius08, AlfredWalsh, AliShaikh85, Alirezazzz, Alisha.4m, Allanvs, Allpower, Amitn, Andres, AndrewHowse, Andros 1337, Andyabides, Anir1uph, Anole 418, AnonMoos, Anshuln95, Antandrus, Antoncampos, Anubhavsharmaa, Apalsola, ApnAEA, Apparition11, Arch dude, Ariele, Armando, Artem-S-Tashkinov, Arthena, Arungm29, Arunsingh16, Asanka000, Ashmoo, Ashutosh.manager007, Atanasov, Avoided, Axeman89, BLAZEXBOY, BaSH PR0MPT, Backwalker, Badr55, Barek, Batbayarl, Beaker1306, Beao, Bearcat, Beetstra, Bender235, Benhocking, BernardZ, Bfaabaa, Big Brother 1984, BigT333, Billyboy1980, Blahma, Blake-, Blakegripling ph, Blobglob, Bloggert, Bluezy, Bmannaa, Bobblewik, Bobo192, Bogdangiusca, Bongomatic, Bongwarrior, Boomshadow, BorgHunter, BorisxXx, BoyanSyarov, Brennish, BrightStarSky, Brockert, BrokenSegue, Brokestudent007, Bruce1ee, BunnyT, C. A. Russell, C.Fred, C311u1ar, C628, CONFIQ, Cablecord, Cameron Scott, Can't sleep, clown will eat me, CanadianLinuxUser, Canterbury Tail, Capricorn42, Catiana 465, CecilWard, Cerebrith, Chamal N, Cheezy man, Cherkash, Cheung1303, Choptube, Chris the speller, ChrisHodgesUK, Chrisissocool, Chronulator, Clarkedexter, ClementSeveillac, CliffC, Clipmode, Closedmouth, Cmreditor, Cntras, Codey123, Coguar, Colibri37, Conversion script, Coolbull20, Coolslko, Cpl Syx, Crevox, Cryonic07, Crysb, Cybercobra, DBigXray, Dale Arnett, Damiens.rf, Damirgraffiti, Dancter, Daniel*D, Daniel575, DanielCD, Danim, Danio, DarkSaber2k, Darth Panda, Davidbspalding, Dazman 1988, Dck7777, Dearsina, Debresser, Delta avi delta, Den fjättrade ankan, Desaivishal14, Desbiadi, Diasimon2003, Dicklyon, Dicostathomas, Dima1, Discospinster, Dissolve, Diyar se, Djr xi, DmitTrix, DocendoDiscimus, Doctorcasey, DomQ, Dostal, Dostick, Download, Drewt, Drpickem, Dsavi.x4, Dyl, E Pluribus Anthony, E000xm, ES Vic, EWikist, Ed g2s, Editor182, Edsuom, Educatednawab, Edward, Eekerz, Eenu, Elfguy, Eliz81, Emmalewis1, Enemenemu, Epoxed, Eraserhead1, Erunestian, Esebi95, Estoy Aquí, Eta 94, Etincelles, Eurocanna, Everyking, FR Soliloquy, Falcon8765, Falconoffrance63, Faramir1138, Feezo, Felix Dance, Fffaaattt, Fieldday-sunday, Fixer88, FleetCommand, Flewis, Flora, Florentino floro, Florin92, Flrn, Flyguy649, Folksong, Fooishbar, Force39, Fraggle81, Fram, France64160, Franz-kafka, Fredrik, FreplySpang, Frymaster, FunkyDuffy, GMRE, Gachen, Galeon54, Galoubet, Gambit 28, Gandhietami, Gareth Griffith-Jones, Gaurav13dubey, Gavinio, Gdo01, Georgy90, Gethresh, Ghodannywahyudi99, Gimboid13, Gnangarra, Gnuton, Gobonobo, Godospoons, Gogo Dodo, Golemeye, Gr1st, Grafen, Graham87, Gratom, GreenJellybean, GreenJoe, Greenshed, Gregorydavid, Gringer, Groshna, Gsarwa, Guaka, Gump Stump, Gunglewack, Gunmetal Angel, Guy M, H.lloyd, H0dd0ck, Haakon, Hadal, Hankwang, Harishua, Harmi banik111, Harriv, HartzR, Hatesonyericsson, Hauskalainen, Havarhen, Hdt83, Head, Hell9, HelloAnnyong, Hemanshu, Heron, Herr Beethoven, HkCaGu, Hmains, Hooperbloob, Hotwiki, Howardchu, Htanna, Hu12, Hustedcarl, Hydrargyrum, Hylene, IJK Principle, Icewindfiresnow, Icseaturtles, Ilovemymac, Imgaril, Immunmotbluescreen, Imperi, Inspirito, Intelligentsium, Interframe, Inzy, Iohannes Animosus, Iphon, Irishnbears, Irstu, Isfisk, Ithinkicrappedmyself, Ixfd64, J. Sketter, J.delanoy, J36miles, JAAqqO, JCDenton2052, JForget, JGRIFF47, JIP, JLaTondre, JNW, Ja 62, Jackollie, Jak123, Jamcib, Jankratochvil, JayceAndTheNews, Jbsegal, Jclemens, Jdl18, Jeffrey Mall, Jeltz, Jensbn, Jeronimo, Jerry, Jerryseinfeld, Jim, Jimp, JmeSaunders, Jmh, Jni, JoeHinks, Joel7687, John, John Appleseed, JonForst6, Jonathan Hall, Jonik, Jopo, JorgeGG, Joseph Solis in Australia, Jovianeye, Jpk, Jrdioko, Jsysinc, JudasJesus, Julesd, Justin Steele, KC., KFP, KUsam, Kabsingh, Kai Ojima, Kalivd, Kallemax, Kalpesh.v.mistry, Kangaroopower, Katous1978, Kcandrsn, Keegan, Keonne, Ketchup, Khanri01, Kickus, Kirosana, Kjetilho, Kjramesh, Kkm010, Klilidiplomus, KoastalBenefitPromo, Konakalla anurag, Konrad Foerstner, Kosyoboy, Krellis, Kukulcan 7560, Kuponadam, Kwamikagami, LG4761, LLentil, Lafraia, Lamat, Larry laptop, LarryGilbert, Ld100, Leandrod, LeaveSleaves, Lectonar, Leo-loErlahell, Lepensky, Levineps, Lewys, Lfh, Lhotkami, Liamgilmartin, Lightmouse, LilHelpa, LittleDan, Loigenth, Lollomama, Look2See1, Loren.wilton, Lotje, Lovesameer9812, Lowflyingowl, Luigiacruz, Luisvmejia, Lukobe, M3lm4tt, MER-C, MMuzammils, Ma.abilash, Mabdul, Magioladitis, Makedonec28, Makeemlighter, Malhonen, Mani1, Manop, Marc Lacoste, Mardus, Marek69, Mark, Mark Arsten, Marmzok, Martarius, Martin.uecker, Mass09, Mastermind 147, Matrobriva, Matt povey, Mattbr, Mav, Maxim4o, Mayhaymate, Mayurg, McSly, McWika, Megahmad, Menphrad, Mert2000, MetroStar, Mhkay, Mic, Michaelbarr123, Microtony, Mike Rosoft, Mimsie, Minimac, Miraceti, Mirmo!, Misspsyb2, MithrandirAgain, Mkidson, Mmsteelers, Mojei, MominS, Momirt, MonaNL, Monkeynoze, Mowsbury, Mr Stephen, Mr. Met 13, MrFawwaz, MrOllie, MrZoolook, MrsAmethystWay, Mumble45, Murph146, Murphy418, Mysdaao, NKDurrani, Nagytibi, Nakon, Nantasatria, Narcisso, NawlinWiki, NeilN, Neilgravir, Neon white, Newone, Newtechpulse, Nielswik, Nikopolis1912, Ninja5624, Ninjustic, NokiaF, Nokiatrader, Noraft, Norden83, Nscheffey, Nubbly, Nubiatech, Number29, Nuno Tavares, Nuttycoconut, Nwpl, Obli, Ohnoitsjamie, OkakiMCMLXX, Oki putera, Oleksandr Kononenko, Oli Filth, Olivier, Omernos, One, Onlynokia, Onorem, Oo64eva, Ooza, Opraco, Ostralek, Otto2011, Oxymoron83, P0ppe, PTSE, Pablo-flores, Palefire, Parthrana, Pascal.Tesson, Pavel Vozenilek, PeeJay2K3, Pengo, Perohanych, Persia2, Pessi, PeteS, Peter Chastain, Petri Krohn, Pgan002, Pgk, Philip Trueman, Phonefinder2007, Pink Bull, Pista235, Pkadam, PkerUNO, Ploca12, Plokijnu, Pmggp, Pne, Pol098, Pony1142, Ppntori, Prakash.Akshat87, Prari, ProhibitOnions, Prokopov, Prolog, Proudfoot 001, Pseudomonas, Pterre, Puckly, Pudeo, Puneeeetjain, Pupster21, Quackdave, Quattrope, QuiteUnusual, Qwyrxian, Qxz, R'n'B, RTG, RVJ, Radagast83, RadicalBender, Ragityman, Rangoon11, Ratinator, Ratul655, Ravi vadgama, Ravo492, Rcawsey, RedWolf, Reliancepowercoin, Rent A Troop, Retired user 0001, Rettetast, Reuvengrish, RexNL, Rhobite, Richard Harvey, Rickt herazor, Riklear, Rjwilmsi, Rklawton, Roamataa, Rob Lindsey, Rob1974, Ron Ritzman, Ronnotel, Rrjanbiah, Rune X2, Ruronimomo, Ruslan0202, Rwalker, S h i v a (Visnu), S3000, SMC, SMP, ST47, Sadunlove, Sai2020, Saimhe, Salilm, Salome5764, Sam Hocevar, Samjuise, Samsara, Sander Säde, Sanixia, Sanjiv swarup, Satchmo2010, Savh, Savvo, Scootey, ScottSteiner, Scupplefish, Seanmilloy, Sebastian Shaw 449, Secaundis, Sedathut, Selimbey, Seqsea, Sfan00 IMG, Sfmammamia, Shadowjams, Sharcho, Shashankbhat, Shaun680, Shawnnicholsonca, Shd, Sherool, Shikker, Shizane, Siafu, SidP, Silpol, Silvah 87, SimonCrowley, SimonThird, Simone, Sinigagl, Skittle, Skyezx, Slavon37, Slo-mo, SmartFace44, SnappingTurtle, Snowolf, Sokratees9, Someguy1221, Soupyjnr, Spangineer, SpigotMap, Squash Racket, Squirtypants, Starblind, Steel, Stephan Leeds, Stephenb, SteveDay, SteveSims, Studioghiblitotoro, Suhailpeerbhai, Suomi Finland 2009, SuperHamster, Sven Manguard, Svgalbertian, Swctg, Syniq, T.O. Rainy Day, THEunique, TYelliot, Tagishsimon, TarzanASG, Tascha96, Tbhotch, Tda, Tedats, Teksosyete, Tellarin, Templetongore, Teqhed, Term061, Test2010, Thaliadrogna, The Man in Question, The Person Who Is Strange, The Random Editor, The Rogue Penguin, The Thing That Should Not Be, TheBlueKnight, TheGreenFaerae, TheYmode, Thebluebeast, Themfromspace, Thingg, Thorvy, Thule, Thunderbrand, TicketMan, Tide rolls, TigerK 69, Timo Honkasalo, Timonoko, Timster69, Tkynerd, Toehead2001, TomB123, Tomisti, Tompagenet, Tomwalden, Tonius, Tony1, Topperfalkon, Tpbradbury, Trakesht, Treschupetes, Tri400, Triage, Trisreed, Tsungik, Tuliopa, Turnstep, Twid, Ufinne, Ulric1313, Ultraviolet scissor flame, Ulysses, Uncle Dick, Unclealex, UpBeat, UpstateNYer, Valentine McKee, Varundbest10, Velella, Venus 9274, Versageek, Vespristiano, Vianello, Vina, Vinayakgole, Vinaywin7, Vininche, Vitund, Vkem, Volgar, Vrenator, Vuo, WIMYV, WJetChao, Wahgujarat, Waycool27, Weatherwax, Welovedoves, Weyes, Wfaulk, Wibbble, Wiki Wonda, Wikipelli, Wikizard2010, Wikizeta, William M. Connolley, WilliamKF, Wimt, WiseOne76, XLerate, Xhienne, Xyzahirm, Yakudza, Yamaaan, Yamamoto Ichiro, Yandman, Yanksox, YellowMonkey, Yousaf465, Yuhani, Yvresgyros, Z10x, ZZninepluralZalpha, Zackwee, Zeno Gantner, Zidonuke, ZirconiumTwice, Zodiak 887, Zpetro, Zsinj, Zunils, امران مجد, محمد المحارب, 55דודו, ஜெலெசின்கி லொந்தூ, □□□□ . □□□ , 1421 anonymous edits

Touchscreen *Source*: http://en.wikipedia.org/w/index.php?title=Touchscreen *Contributors*: 00 diver ice, 041744, 0612, 12J34O56E78, 16@r, 2A01:E35:2E7B:4E50:4012:54FF:FE8D:47FB, 42deepthought, 5 albert square, A. B., A. S. Castanza, A.Ward, A876, ABF, AGToth, Abhishekmathur, Acalamari, Acdx, ActPerspective, Adot, Adrian J. Hunter, Adultswimla, Aerosol monkey, Aetson Eo, Ahoerstemeier, AlainV, Alansohn, Alexhatesmil, Alpabarot, Aman s shah, Amberrock, Angelastic, Angr, Anon9695, Anonymous101, Anthony Appleyard, Appraiser, Arny, Arthena, Atomicblah, BD2412, Back ache, Badon, Barek, Beano, Beetstra, Belmond, Belugaperson, Bender235, Benjherb, Bennetto, BernardSumption, Bibinboi, Big2thumbs, Bigblok403, Bigpresh, Bill Cannon, Bill F Evans, Blueheel, Bobo192, Bongwarrior, Branin, Brotherterrance, Bryan Derksen, Cab88, Cadiomals, Calabe1992, CalumCook234, CanadianLinuxUser, Capricorn42, Casimirpo, Cbessonleaud, Ccook, Ceranthor, CesarB, CezarkennySeF, Chanlord, CharlesC, CherryHintonBlue, Chillllls, Chip123456, Chris3145, Chrisbolt, Chriswiki, Chriswilkins2, Ckatz, Clara Anthony, Codeczero, Colenso, ColinHelvensteijn, Cometstyles, CommonsDelinker, Communisthamster, Coolbho3000, Coolhawks88, Corryn, Corvus cornix, Courcelles, Cp111, Crazy Chall, Cryptex, Cvachon, Cy21, D39, DBZROCKS, DMahalko, DVdm, Damnreds, Dan Harkless, Dancter, DanielLC, Darlo888, David Latapie, Dawnseeker2000, Dcirovic, Deficit, Degdol, Delphii, Demenne, Dfiedor, Dicklyon, Discospinster, DocWatson42, Dominik.krejcik, Donmu, Drewnoakes, Dtgriffith, Dtgriscom, Dwaynedomi, Easitechcn, Eddau, EdwardHades, EdwardZhao, Eliz81, Enigmaticland, Enoril, Epbr123, Erianna, Esatyilmaz, Everyking, FarahAvi, Farhat, Fetze, Fra 011 011, Fraggle81, Frap, Fred Bradstadt, Fresheneesz, Fullrabb, Fullstop, Funandtrvl, Funnyfarmofdoom, GB fan, GCFreak2, GT5162, Gaius Cornelius, Gblindmann, Gdcox, Geekstuff, GeneMosher, Genius101, Georgy90, Gg4rest, Giftlite, Gmaxwell, Gnorthup, Golgofrinchian, Graham87, Gripeape, Gsarwa, Guy Harris, Haakon, HaeB, Haikallp, Hbdragon88, Hede2000, Heron, Hgb asicwizard, Horshamknowsbest, Hqm2009, INkubusse, Ingolfson, Ioeth, Ionutpopa, Iridescent, J.delanoy, JD554, JDP90, JJ Harrison, JJJJust, JaGa, Jackarooneyo, Jagged 85, Jamesooders, Jamuraa, Jdk06734, Jeff G., JeremyA, Jerome Charles Potts, Jgoldfar, Jidanni, Jmabel, Jmrowland, Jncraton, Joeybabe, Johnflux, Jojhutton, Jondel, Jrockley, Julesd, Jumping cheese, Jvr725, KARTHIK PRAVEEN KUMAR, KaiAdin, Kateshortforbob, Kerrick Staley, Kevin chen2003, Khukri, Kinema, Kingpin13, Kingturtle, Kipholbeck, Kitsunegami, Kittman11, Klubneeka, KnowledgeOfSelf, Kostmo, Kraftlos, Kylemcinnes, Kyng, L Kensington, L2blackbelt, Lambiam, Lazylaces, Lester, Lhasapso, Liftarn, Lilcamel, LimoWreck, Lincoln Griffiths, Lradrama, Lrdwhyt, Lucasrenzi, LuoShengli, Lzur, MER-C, Mahaboyd, Marasama, Marcus Qwertyus, Marek69, Mark Arsten, Martarius, Materialscientist, Matty, Mavericks12, MaxPont, Maximo rivi, Maxí, Medvedev, Melack, Menchi, Mentifisto, MetaManFromTomorrow, Metageek, Michael Hardy, Mikael Häggström, Mikeblas, Mindmatrix, Mogism, Moletrouser, Mongo hughes, MoraSique, Mosmof, Moxfyre, MrOllie, Mschlindwein, Mtnman79, MyNameWasTaken, Mygerardromance, Naar, Nanoamp, Nathannich, NawlinWiki, Neffk, Neha27, Nick mitu 25, NickBush24, Nicknick09, Nikevich, Nmh, Noeatingallowed, Nono64, Nopetro, Nortonjb, Nudecline, Nuttycoconut, Oaragones, Oball, Oda Mari, Omegatron, Ost316, Pamlovely, Parsiferon, Paul A, Paxsimius, PenComputingPerson, PeterMottola, Petrb, Pgk, Phantomsteve, PhilKnight, Philip Trueman, Philspivey, Physboy, Piano non troppo, Pinky9292, Platdujour, Pmarshal, Poiuyt Man, Postrach, Puchiko, Pyrilium, R'n'B, RSPINC, RW Marloe, RayquazaDialgaWeird2210, RazorICE, Rchiao, Reaper Eternal, Redlioness, RememberMe?, RichardBronosky, Richiekim, Ritz, Riverarvi, Rjwilmsi, RobertCailliau, Roflpedia, Roux, Roxtone, Rspeed, RyanGerbil10, SCEhardt, SF007, Saddhiyama, Saebjorn, Sahils1512, Salvar, SchfiftyThree, SeanGustafson, SebastianHelm, Senator Palpatine, Sendai2ci, Seniorfrogs, Shalom Yechiel, Shawn in Montreal, Shoez, Sietse Snel, Sigmundur, Signalhead, Silver Edge, Skarnani, Sl24056, Smartyhall, Special Cases, Speer320, Staffwaterboy, Steve Quinn, Steven Walling, Stroppolo, T.A Shirakawa, TJ Spyke, TastyPoutine, TeemuN, Thaddius, The Anome, TheNewPhobia, Thealimighty1, Thecurran91, Themfromspace, Themoonisdown09, Thetay24, Thingg, Thumperward, Tide rolls, Tills, Tktktk, Tobias Bergemann, Todd Vierling, Toddst1, Tombrabbin, Tomcully, Tomj, Tomtom1540, Torchiest, Tounitor, Toussaint, Toytown Mafia, Traceywaldron, Trane X, Traxs7, Tsuba, Tt 225, Tursic, TycoElo, Tyro, Ulric1313, Underpants, Urness.sam, Utcursch, Ute in DC, Uvlr, Vanished user 34958, Vgautham 91, Virtualsets, Vkojouharov, Vpltd, Wallen-zhou, Warhammer100, Wasbuxton, Wavelength, Wbm1058, Welsh, Whatifound, Whoop whoop, WikHead, Wiki13, Wikiscool, Wimt, Wingedsubmariner, Wirrad, Woohookitty, Xtophyr Wright, Yermyahu, Yerpo, YolanCh, Yudiweb, Zachpw, Zazpot, Zbblanton, Zhu8, Zollerriia, Zsc100, Zythe, ماني, अीष भटनागर, 986 anonymous edits

Series_40 *Source*: http://en.wikipedia.org/w/index.php?title=Series_40 *Contributors*: 1exec1, Agreimann, Alex Perry, Aniac, Ashrutisingh, Azuya, Bjelleklang, Blackvienna, Blakegripling ph, C933103, Calltech, Canaima, Centrx, Chapzboy, Charivari, Cnj, Daverocks, Davidbengtenglund, Diwas, Djlarz, Djr xi, Dktz, Dublinclontarf, Elspif, Ettrig, Evilboy, FleetCommand, Flod logic,

Florin92, GRighta, Gaius Cornelius, Gaming&Computing, Hjorten, Imroy, ItsMeOrYou, JLaTondre, Jarekadam, Jasonon, Jesjimher, Joneilim, LordBen5000, MMuzammils, Mburdis, Mdwh, Milan292208, Monkeyhousetim, Msulik, Munamankeli, Park3r, Petri Krohn, Polisher of Cobwebs, Polluks, Rich Farmbrough, Rlobkovsky, Roleplayer, Rudiedude, Rwwww, SimonMenashy, Sofoklas, The Seventh Taylor, TheParanoidOne, Tirkfl, Txuspe, UKER, Vegaswikian, W like wiki, We hope, Xkury, Zakawer, ZeroOne, Zhernovoi, 106 anonymous edits

Graphics_display_resolution *Source*: http://en.wikipedia.org/w/index.php?title=Graphics_display_resolution *Contributors*: 0612, 123GhostMonkey, AGiorgio08, Acroterion, Allens, Anders Feder, Arvindtm, Asiflge, Audriusa, Auntof6, Autopilot, Beeblebroxisaloser, Bender235, Bomazi, Bongwarrior, Bubba73, Confession0791, Cool like ice, Crissov, Cxp3, DGG, Dan198792, Deadly Coordinates, Deliriousbb, DocWatson42, DragonHawk, Edward, Elk Salmon, Emmertex, Faizanalivarya, Fatphil, FlamingGirrafe, Flghtmstr1, Fluffylouis, GerbilSoft, Giraffedata, Glirio, GrandDrake, Hamez0, Hell Pé, Hellbus, Hippo99, HotXRock, Hydrox, Indrek, Jamie367p, Jchen8809, Jeff Carr, Jerryobject, JohnHorak, JonathanSMB, Ju1i3t, Love2scoot, M.O.X, MN, Makyen, Mhickman78, Mishler77, Misty MH, Morenus, NaN135709, Nazar, Neuhaus, Nivlak7, Oilosiso, Oliver H, Phluid61, Q Chris, R'n'B, Ramesh Ramaiah, Ringbang, Rursus, Rz-de, Sainath468, Samlikeswiki, Sonjaaa, Spryng, TMV943, TheJJJunk, Thenewstinks, Theopolisme, Trevj, Trusilver, Txaggiemichael, UU, Vigga, Wtshymanski, Yashlondhe, Yasuna, YolentaShield, Ysangkok, Yz39, Zazsasha, 327 anonymous edits

Java_Platform,_Micro_Edition *Source*: http://en.wikipedia.org/w/index.php?title=Java_Platform%2C_Micro_Edition *Contributors*: .:Ajvol:., Aidan W, Alainr345, Allstarecho, Amareto2, Anandviveksatapathi, AnonMoos, Arosa, Asiftasleem, Aspro, Benjaminhill, Blablablob, Blowdart, Bomazi, Brownb2, Bryan Derksen, Bt227, Casperl, Centrx, CesarB, Chaiken, Chekristo, Chester Markel, Chowbok, Chris857, Cilitate, Codetiger, DStoykov, Damian Yerrick, Danrah, Dave Nelson, David Edgar, David Haslam, David Johnson, Dcoetzee, Deineka, Doug Bell, Eahiv, Ehudshapira, Eliashedberg, Enchanter, Enough, Eortiz, Eskalin, Explanator, Face, Faisal.akeel, Flamurai, Frigotoni, Gadfium, GaussTek, GosiaCh, Gsmdev, Harryboyles, Hervegirod, ICEAGE, IJK Principle, Idlenism, Ilaiho, Intgr, Ivant, J Di, J.delanoy, J2meed, Jeffq, Jgrovert, Jjaazz, Jlin, Jmreinhart, Jpta, Kavermei, Kishonti, Kkm010, Krischik, L888Y5, Lainestl, Lbecque, LegalEagle2, Leon de Beer, Magister Mathematicae, Martarius, Mathiastck, MattiPaavola, Melab-1, Mieciu K, Mithaca, Mjwrite, Mlk, MoreNet, Nakon, Nandssiib, NapoliRoma, Nealmcb, Neilc, Nephtes, Nissir, Nixdorf, Nixeagle, Nopetro, Northgrove, NotWith, Npovmachine, OS2Warp, Ohnoitsjamie, Oli Filth, Oolong, Palopt, Paul Poutanen, Peshwali, Pit, Pnm, Pok.lee, Posix memalign, Pwjb, QTCaptain, Rablwupei, Rafaelwanjiku, Reffik, Ringbang, RomanGrigoryev, SD5, Seanhan, Seeknshare, Sfmontyo, Shizhao, SirFozzie, Skj.saurabh, Slady, SlubGlub, SmackEater, Suruena, Tarquin, Ted nw, TedBaker88, Thanhbv, Thumperward, Tintinobelisk, Tokek, Tomsoft, Topbanana, Traroth, Treutwein, Tzartzam, Udittmer, Ufim, UnitedStatesian, Uzume, Vegaswikian, Whitepaw, WikiFanCMC, Winterst, Wmahan, Wordcraft3, Yahlowgrin, Zfr, Zidane2k1, ZimZalaBim, Zumbo, 274 anonymous edits

Image Sources, Licenses and Contributors

File:Nokia wordmark.svg *Source*: http://en.wikipedia.org/w/index.php?title=File:Nokia_wordmark.svg *License*: Trademarked *Contributors*: Apalsola, Bencmq, ELeschev, Editor182, Hautala, Yarl, 1 anonymous edits

File:Decrease2.svg *Source*: http://en.wikipedia.org/w/index.php?title=File:Decrease2.svg *License*: Public Domain *Contributors*: Sarang

Image:Fredrik Idestam.png *Source*: http://en.wikipedia.org/w/index.php?title=File:Fredrik_Idestam.png *License*: Public Domain *Contributors*: -Majestic-, Apalsola, Jpk, Martin H.

Image:Leo Mechelin (cropped).png *Source*: http://en.wikipedia.org/w/index.php?title=File:Leo_Mechelin_(cropped).png *License*: Public Domain *Contributors*: -Majestic-, Martin H.

File:Nokia 150 and nokia 1100.jpg *Source*: http://en.wikipedia.org/w/index.php?title=File:Nokia_150_and_nokia_1100.jpg *License*: Public Domain *Contributors*: Lvova Anastasiya (Львова Анастасия, Lvova)

File:Nokia booklet 3g-10 (3949263497).jpg *Source*: http://en.wikipedia.org/w/index.php?title=File:Nokia_booklet_3g-10_(3949263497).jpg *License*: Creative Commons Attribution-Sharealike 3.0 *Contributors*: http://thenokiablog.com/Reposted by Mark Guim from United States

File:Nokia HQ.jpg *Source*: http://en.wikipedia.org/w/index.php?title=File:Nokia_HQ.jpg *License*: Creative Commons Attribution-Sharealike 3.0,2.5,2.0,1.0 *Contributors*: -Majestic-

File:Nokia evolucion tamaño.jpg *Source*: http://en.wikipedia.org/w/index.php?title=File:Nokia_evolucion_tamaño.jpg *License*: Public Domain *Contributors*: Jorge Barrios

File:All 9xxx.png *Source*: http://en.wikipedia.org/w/index.php?title=File:All_9xxx.png *License*: GNU Free Documentation License *Contributors*: derivative work: -Majestic- (talk) All9xxx.jpg: Original uploader was R@y at de.wikipedia (2004-04-03)

File:Nokia N8 (front view).jpg *Source*: http://en.wikipedia.org/w/index.php?title=File:Nokia_N8_(front_view).jpg *License*: Public Domain *Contributors*: Editor182, Vivinnl, X-Pilot

File:SymbianWMWP7USMarketShare.png *Source*: http://en.wikipedia.org/w/index.php?title=File:SymbianWMWP7USMarketShare.png *License*: Creative Commons Attribution-Sharealike 3.0 *Contributors*: User:Enemenemu

File:Nokia E55 01.jpg *Source*: http://en.wikipedia.org/w/index.php?title=File:Nokia_E55_01.jpg *License*: Creative Commons Attribution-Sharealike 2.0 *Contributors*: James Nash

File:Nokia N900-1.jpg *Source*: http://en.wikipedia.org/w/index.php?title=File:Nokia_N900-1.jpg *License*: Creative Commons Attribution-Sharealike 3.0,2.5,2.0,1.0 *Contributors*: User:Ilya Voyager

File:Nokia E90 communicator.JPG *Source*: http://en.wikipedia.org/w/index.php?title=File:Nokia_E90_communicator.JPG *License*: Creative Commons Attribution 3.0 *Contributors*: Georgy90

File:Nokia5800xpress.png *Source*: http://en.wikipedia.org/w/index.php?title=File:Nokia5800xpress.png *License*: Creative Commons Attribution-Sharealike 3.0 *Contributors*: Nokia_5800_XpressMusic_Browser.jpg: Jupter-manzana derivative work: TheAdam0s (talk)

File:Flag of Canada.svg *Source*: http://en.wikipedia.org/w/index.php?title=File:Flag_of_Canada.svg *License*: Public Domain *Contributors*: Anomie

File:Flag of Finland.svg *Source*: http://en.wikipedia.org/w/index.php?title=File:Flag_of_Finland.svg *License*: Public Domain *Contributors*: Drawn by User:SKopp

File:Flag of the United States.svg *Source*: http://en.wikipedia.org/w/index.php?title=File:Flag_of_the_United_States.svg *License*: Public Domain *Contributors*: Anomie

File:Flag of Switzerland.svg *Source*: http://en.wikipedia.org/w/index.php?title=File:Flag_of_Switzerland.svg *License*: Public Domain *Contributors*: User:Marc Mongenet Credits: User:-xfi- User:Zscout370

File:Flag of Germany.svg *Source*: http://en.wikipedia.org/w/index.php?title=File:Flag_of_Germany.svg *License*: Public Domain *Contributors*: Anomie

File:Flag of Norway.svg *Source*: http://en.wikipedia.org/w/index.php?title=File:Flag_of_Norway.svg *License*: Public Domain *Contributors*: Dbenbenn

File:Flag of France.svg *Source*: http://en.wikipedia.org/w/index.php?title=File:Flag_of_France.svg *License*: Public Domain *Contributors*: Anomie

File:Nokia Connecting People.svg *Source*: http://en.wikipedia.org/w/index.php?title=File:Nokia_Connecting_People.svg *License*: Public Domain *Contributors*: Original uploader was -Majestic- at en.wikipedia

File:Nokian logo.svg *Source*: http://en.wikipedia.org/w/index.php?title=File:Nokian_logo.svg *License*: Public Domain *Contributors*: Z10x at en.wikipedia

File:Nokian pääkonttori Keilaniemessä.jpg *Source*: http://en.wikipedia.org/w/index.php?title=File:Nokian_pääkonttori_Keilaniemessä.jpg *License*: GNU Free Documentation License *Contributors*: J-P Kärnä

Image:Touchscreen IMG 2796.jpg *Source*: http://en.wikipedia.org/w/index.php?title=File:Touchscreen_IMG_2796.jpg *License*: Creative Commons Attribution-Sharealike 3.0 *Contributors*: User:Eddau

File:IPad 2 on stand.jpg *Source*: http://en.wikipedia.org/w/index.php?title=File:IPad_2_on_stand.jpg *License*: Creative Commons Attribution 2.0 *Contributors*: Matthew Pearce

file:CERN-Stumpe Capacitance Touchscreen.jpg *Source*: http://en.wikipedia.org/w/index.php?title=File:CERN-Stumpe_Capacitance_Touchscreen.jpg *License*: Creative Commons Attribution *Contributors*: Maximilien Brice

File:ViewTouch1986.jpg *Source*: http://en.wikipedia.org/w/index.php?title=File:ViewTouch1986.jpg *License*: Creative Commons Attribution-Sharealike 3.0 *Contributors*: User:GeneMosher

File:Capacitive touchscreen.jpg *Source*: http://en.wikipedia.org/w/index.php?title=File:Capacitive_touchscreen.jpg *License*: Creative Commons Attribution 3.0 *Contributors*: Medvedev

File:PCT Globe.jpg *Source*: http://en.wikipedia.org/w/index.php?title=File:PCT_Globe.jpg *License*: Creative Commons Zero *Contributors*: Gblindmann

Image:Platovterm1981.jpg *Source*: http://en.wikipedia.org/w/index.php?title=File:Platovterm1981.jpg *License*: Creative Commons Attribution 3.0 *Contributors*: Mtnman79

File:Pointed nail.png *Source*: http://en.wikipedia.org/w/index.php?title=File:Pointed_nail.png *License*: Public Domain *Contributors*: Mikael Häggström

File:Touchscreen protector.jpg *Source*: http://en.wikipedia.org/w/index.php?title=File:Touchscreen_protector.jpg *License*: Creative Commons Attribution-Sharealike 3.0 *Contributors*: User:Riverarvi

Image:Nokia 6300.jpg *Source*: http://en.wikipedia.org/w/index.php?title=File:Nokia_6300.jpg *License*: Public Domain *Contributors*: Original uploader was Racklever at en.wikipedia

File:Symbian S40 v10.80.png *Source*: http://en.wikipedia.org/w/index.php?title=File:Symbian_S40_v10.80.png *License*: Creative Commons Attribution-Sharealike 3.0 *Contributors*: Blackvienna, 1 anonymous edits

Image:Vector Video Standards2.svg *Source*: http://en.wikipedia.org/w/index.php?title=File:Vector_Video_Standards2.svg *License*: GNU Free Documentation License *Contributors*: Original uploader was XXV at en.wikipedia Later version(s) were uploaded by Jjalocha, Aihtdikh at en.wikipedia.

Image:Qvga.svg *Source*: http://en.wikipedia.org/w/index.php?title=File:Qvga.svg *License*: Public Domain *Contributors*: Stephantom

File:IBM T221.jpg *Source*: http://en.wikipedia.org/w/index.php?title=File:IBM_T221.jpg *License*: Creative Commons Attribution-Sharealike 3.0 *Contributors*: User:Autopilot

GNU Free Documentation License Version 1.2, November 2002 Copyright (C) 2000,2001,2002 Free Software Foundation, Inc. 59 Temple Place, Suite 330, Boston, MA 02111-1307 USA Everyone is permitted to copy and distribute verbatim copies of this license document, but changing it is not allowed.

0. PREAMBLE

The purpose of this License is to make a manual, textbook, or other functional and useful document "free" in the sense of freedom: to assure everyone the effective freedom to copy and redistribute it, with or without modifying it, either commercially or noncommercially. Secondarily, this License preserves for the author and publisher a way to get credit for their work, while not being considered responsible for modifications made by others. This License is a kind of "copyleft", which means that derivative works of the document must themselves be free in the same sense. It complements the GNU General Public License, which is a copyleft license designed for free software. We have designed this License in order to use it for manuals for free software, because free software needs free documentation: a free program should come with manuals providing the same freedoms that the software does. But this License is not limited to software manuals; it can be used for any textual work, regardless of subject matter or whether it is published as a printed book. We recommend this License principally for works whose purpose is instruction or reference.

1. APPLICABILITY AND DEFINITIONS

This License applies to any manual or other work, in any medium, that contains a notice placed by the copyright holder saying it can be distributed under the terms of this License. Such a notice grants a world-wide, royalty-free license, unlimited in duration, to use that work under the conditions stated herein. The "Document", below, refers to any such manual or work. Any member of the public is a licensee, and is addressed as "you". You accept the license if you copy, modify or distribute the work in a way requiring permission under copyright law. A "Modified Version" of the Document means any work containing the Document or a portion of it, either copied verbatim, or with modifications and/or translated into another language. A "Secondary Section" is a named appendix or a front-matter section of the Document that deals exclusively with the relationship of the publishers or authors of the Document to the Document's overall subject (or to related matters) and contains nothing that could fall directly within that overall subject. (Thus, if the Document is in part a textbook of mathematics, a Secondary Section may not explain any mathematics.) The relationship could be a matter of historical connection with the subject or with related matters, or of legal, commercial, philosophical, ethical or political position regarding them. The "Invariant Sections" are certain Secondary Sections whose titles are designated, as being those of Invariant Sections, in the notice that says that the Document is released under this License. If a section does not fit the above definition of Secondary then it is not allowed to be designated as Invariant. The Document may contain zero Invariant Sections. If the Document does not identify any Invariant Sections then there are none. The "Cover Texts" are certain short passages of text that are listed, as Front-Cover Texts or Back-Cover Texts, in the notice that says that the Document is released under this License. A Front-Cover Text may be at most 5 words, and a Back-Cover Text may be at most 25 words. A "Transparent" copy of the Document means a machine-readable copy, represented in a format whose specification is available to the general public, that is suitable for revising the document straightforwardly with generic text editors or (for images composed of pixels) generic paint programs or (for drawings) some widely available drawing editor, and that is suitable for input to text formatters or for automatic translation to a variety of formats suitable for input to text formatters. A copy made in an otherwise Transparent file format whose markup, or absence of markup, has been arranged to thwart or discourage subsequent modification by readers is not Transparent. An image format is not Transparent if used for any substantial amount of text. A copy that is not "Transparent" is called "Opaque". Examples of suitable formats for Transparent copies include plain ASCII without markup, Texinfo input format, LaTeX input format, SGML or XML using a publicly available DTD, and standard-conforming simple HTML, PostScript or PDF designed for human modification. Examples of transparent image formats include PNG, XCF and JPG. Opaque formats include proprietary formats that can be read and edited only by proprietary word processors, SGML or XML for which the DTD and/or processing tools are not generally available, and the machine-generated HTML, PostScript or PDF produced by some word processors for output purposes only. The "Title Page" means, for a printed book, the title page itself, plus such following pages as are needed to hold, legibly, the material this License requires to appear in the title page. For works in formats which do not have any title page as such, "Title Page" means the text near the most prominent appearance of the work's title, preceding the beginning of the body of the text. A section "Entitled XYZ" means a named subunit of the Document whose title either is precisely XYZ or contains XYZ in parentheses following text that translates XYZ in another language. (Here XYZ stands for a specific section name mentioned below, such as "Acknowledgements", "Dedications", "Endorsements", or "History".) To "Preserve the Title" of such a section when you modify the Document means that it remains a section "Entitled XYZ" according to this definition. The Document may include Warranty Disclaimers next to the notice which states that this License applies to the Document. These Warranty Disclaimers are considered to be included by reference in this License, but only as regards disclaiming warranties: any other implication that these Warranty Disclaimers may have is void and has no effect on the meaning of this License.

2. VERBATIM COPYING

You may copy and distribute the Document in any medium, either commercially or noncommercially, provided that this License, the copyright notices, and the license notice saying this License applies to the Document are reproduced in all copies, and that you add no other conditions whatsoever to those of this License. You may not use technical measures to obstruct or control the reading or further copying of the copies you make or distribute. However, you may accept compensation in exchange for copies. If you distribute a large enough number of copies you must also follow the conditions in section 3. You may also lend copies, under the same conditions stated above, and you may publicly display copies.

3. COPYING IN QUANTITY

If you publish printed copies (or copies in media that commonly have printed covers) of the Document, numbering more than 100, and the Document's license notice requires Cover Texts, you must enclose the copies in covers that carry, clearly and legibly, all these Cover Texts: Front-Cover Texts on the front cover, and Back-Cover Texts on the back cover. Both covers must also clearly and legibly identify you as the publisher of these copies. The front cover must present the full title with all words of the title equally prominent and visible. You may add other material on the covers in addition. Copying with changes limited to the covers, as long as they preserve the title of the Document and satisfy these conditions, can be treated as verbatim copying in other respects. If the required texts for either cover are too voluminous to fit legibly, you should put the first ones listed (as many as fit reasonably) on the actual cover, and continue the rest onto adjacent pages. If you publish or distribute Opaque copies of the Document numbering more than 100, you must either include a machine-readable Transparent copy along with each Opaque copy, or state in or with each Opaque copy a computer-network location from which the general network-using public has access to download using public-standard network protocols a complete Transparent copy of the Document, free of added material. If you use the latter option, you must take reasonably prudent steps, when you begin distribution of Opaque copies in quantity, to ensure that this Transparent copy will remain thus accessible at the stated location until at least one year after the last time you distribute an Opaque copy (directly or through your agents or retailers) of that edition to the public. It is requested, but not required, that you contact the authors of the Document well before redistributing any large number of copies, to give them a chance to provide you with an updated version of the Document.

4. MODIFICATIONS

You may copy and distribute a Modified Version of the Document under the conditions of sections 2 and 3 above, provided that you release the Modified Version under precisely this License, with the Modified Version filling the role of the Document, thus licensing distribution and modification of the Modified Version to whoever possesses a copy of it. In addition, you must do these things in the Modified Version: A. Use in the Title Page (and on the covers, if any) a title distinct from that of the Document, and from those of previous versions (which should, if there were any, be listed in the History section of the Document). You may use the same title as a previous version if the original publisher of that version gives permission. B. List on the Title Page, as authors, one or more persons or entities responsible for authorship of the modifications in the Modified Version, together with at least five of the principal authors of the Document (all of its principal authors, if it has fewer than five), unless they release you from this requirement. C. State on the Title page the name of the publisher of the Modified Version, as the publisher. D. Preserve all the copyright notices of the Document. E. Add an appropriate copyright notice for your modifications adjacent to the other copyright notices. F. Include, immediately after the copyright notices, a license notice giving the public permission to use the Modified Version under the terms of this License, in the form shown in the Addendum below. G. Preserve in that license notice the full lists of Invariant Sections and required Cover Texts given in the Document's license notice. H. Include an unaltered copy of this License. I. Preserve the section Entitled "History", Preserve its Title, and add to it an item stating at least the title, year, new authors, and publisher of the Modified Version as given on the Title Page. If there is no section Entitled "History" in the Document, create one stating the title, year, authors, and publisher of the Document as given on its Title Page, then add an item describing the Modified Version as stated in the previous sentence. J. Preserve the network location, if any, given in the Document for public access to a Transparent copy of the Document, and likewise the network locations given in the Document for previous versions it was based on. These may be placed in the "History" section. You may omit a network location for a work that was published at least four years before the Document itself, or if the original publisher of the version it refers to gives permission. K. For any section Entitled "Acknowledgements" or "Dedications", Preserve the Title of the section, and preserve in the section all the substance and tone of each of the contributor acknowledgements and/or dedications given therein. L. Preserve all the Invariant Sections of the Document, unaltered in their text and in their titles. Section numbers or the equivalent are not considered part of the section titles. M. Delete any section Entitled "Endorsements". Such a section may not be included in the Modified Version. N. Do not retitle any existing section to be Entitled "Endorsements" or to conflict in title with any Invariant Section. O. Preserve any Warranty Disclaimers. If the Modified Version includes new front-matter sections or appendices that qualify as Secondary Sections and contain no material copied from the Document, you may at your option designate some or all of these sections as invariant. To do this, add their titles to the list of Invariant Sections in the Modified Version's license notice. These titles must be distinct from any other section titles. You may add a section Entitled "Endorsements", provided it contains nothing but endorsements of your Modified Version by various parties--for example, statements of peer review or that the text has been approved by an organization as the authoritative definition of a standard. You may add a passage of up to five words as a Front-Cover Text, and a passage of up to 25 words as a Back-Cover Text, to the end of the list of Cover Texts in the Modified Version. Only one passage of Front-Cover Text and one of Back-Cover Text may be added by (or through arrangements made by) any one entity. If the Document already includes a cover text for the same cover, previously added by you or by arrangement made by the same entity you are acting on behalf of, you may not add another; but
you may replace the old one, on explicit permission fro
previous publisher that added the old one. The author(s
publisher(s) of the Document do not by this License
permission to use their names for publicity for or to assert or
endorsement of any Modified Version.

5. COMBINING DOCUMENTS

You may combine the Document with other documents rel
under this License, under the terms defined in section 4 abo
modified versions, provided that you include in the combinat
of the Invariant Sections of all of the original docur
unmodified, and list them all as Invariant Sections of
combined work in its license notice, and that you preserve a
Warranty Disclaimers. The combined work need only contai
copy of this License, and multiple identical Invariant Section
be replaced with a single copy. If there are multiple Inv
Sections with the same name but different contents, mak
title of each such section unique by adding at the end of
parentheses, the name of the original author or publisher c
section if known, or else a unique number. Make the
adjustment to the section titles in the list of Invariant Sectic
the license notice of the combined work. In the combinatio
must combine any sections Entitled "History" in the v
original documents, forming one section Entitled "His
likewise combine any sections Entitled "Acknowledgements"
any sections Entitled "Dedications". You must delete all se
Entitled "Endorsements".

6. COLLECTIONS OF DOCUMENTS

You may make a collection consisting of the Document and
documents released under this License, and replace
individual copies of this License in the various documents v
single copy that is included in the collection, provided tha
follow the rules of this License for verbatim copying of each
documents in all other respects. You may extract a
document from such a collection, and distribute it indivi
under this License, provided you insert a copy of this Licens
the extracted document, and follow this License in all
respects regarding verbatim copying of that document.

7. AGGREGATION WITH INDEPENDENT WORKS

A compilation of the Document or its derivatives with
separate and independent documents or works, in or on a v
of a storage or distribution medium, is called an "aggregate"
copyright resulting from the compilation is not used to lim
legal rights of the compilation's users beyond what the indi
works permit. When the Document is included in an aggr
this License does not apply to the other works in the aggr
which are not themselves derivative works of the Document.
Cover Text requirement of section 3 is applicable to these c
of the Document, then if the Document is less than one half
entire aggregate, the Document's Cover Texts may be plac
covers that bracket the Document within the aggregate, c
electronic equivalent of covers if the Document is in elec
form. Otherwise they must appear on printed covers that br
the whole aggregate.

8. TRANSLATION

Translation is considered a kind of modification, so you
distribute translations of the Document under the terms of se
4. Replacing Invariant Sections with translations requires sp
permission from their copyright holders, but you may in
translations of some or all Invariant Sections in addition t
original versions of these Invariant Sections. You may inclu
translation of this License, and all the license notices i
Document, and any Warranty Disclaimers, provided that you
include the original English version of this License and
original versions of those notices and disclaimers. In case
disagreement between the translation and the original versi
this License or a notice or disclaimer, the original versio
prevail. If a section in the Document is Er
"Acknowledgements", "Dedications", or "History", the require
(section 4) to Preserve its Title (section 1) will typically re
changing the actual title.

9. TERMINATION

You may not copy, modify, sublicense, or distribute
Document except as expressly provided for under this Lic
Any other attempt to copy, modify, sublicense or distribut
Document is void, and will automatically terminate your
under this License. However, parties who have received co
or rights, from you under this License will not have their lice
terminated so long as such parties remain in full compliance.

10. FUTURE REVISIONS OF THIS LICENSE

The Free Software Foundation may publish new, re
versions of the GNU Free Documentation License from tin
time. Such new versions will be similar in spirit to the pr
version, but may differ in detail to address new problen
concerns. See http://www.gnu.org/copyleft/. Each version of
License is given a distinguishing version number. If the Docu
specifies that a particular numbered version of this Licens
any later version" applies to it, you have the option of follo
the terms and conditions either of that specified version or o
later version that has been published (not as a draft) by the
Software Foundation. If the Document does not specify a ve
number of this License, you may choose any version
published (not as a draft) by the Free Software Founda
ADDENDUM: How to use this License for your documents To
this License in a document you have written, include a copy c
License in the document and put the following copyright
license notices just after the title page: Copyright (c) YEAR Y
NAME. Permission is granted to copy, distribute and/or m
this document under the terms of the GNU Free Document
License, Version 1.2 or any later version published by the
Software Foundation; with no Invariant Sections, no Front-C
Texts, and no Back-Cover Texts. A copy of the license is inc
in the section entitled "GNU Free Documentation License". I
have Invariant Sections, Front-Cover Texts and Back-C
Texts, replace the "with...Texts." line with this: with the Inva
Sections being LIST THEIR TITLES, with the Front-Cover T
being LIST, and with the Back-Cover Texts being LIST. If
have Invariant Sections without Cover Texts, or some c
combination of the three, merge those two alternatives to su
situation. If your document contains nontrivial example
program code, we recommend releasing these example
parallel under your choice of free software license, such as
GNU General Public License, to permit their use in free softw

CPSIA information can be obtained at www.ICGtesting.com
Printed in the USA
LVOW051643171212

312059LV00004B/650/P

9 786201 901711